The Concise Dictionary of Crime and Justice

Second Edition

The Concise Dictionary of Crime and Justice

Second Edition

Mark S. Davis

Los Angeles | London | New Delhi
Singapore | Washington DC | Boston

Los Angeles | London | New Delhi
Singapore | Washington DC | Boston

FOR INFORMATION:

SAGE Publications, Inc.
2455 Teller Road
Thousand Oaks, California 91320
E-mail: order@sagepub.com

SAGE Publications Ltd.
1 Oliver's Yard
55 City Road
London EC1Y 1SP
United Kingdom

SAGE Publications India Pvt. Ltd.
B 1/I 1 Mohan Cooperative Industrial Area
Mathura Road, New Delhi 110 044
India

SAGE Publications Asia-Pacific Pte. Ltd.
3 Church Street
#10-04 Samsung Hub
Singapore 049483

Printed in the United States of America

Cataloging-in-publication data is available for this title from the Library of Congress.

ISBN 978-1-4833-8093-3

Acquisitions Editor: Jerry Westby
Editorial Assistant: Laura Kirkhuff
Production Editor: Kelly DeRosa
Copy Editor: Janet Ford
Typesetter: C&M Digitals (P) Ltd.
Proofreader: Sue Irwin
Cover Designer: Karine Hovsepian
Marketing Manager: Terra Schultz

Contents

Preface to the New Edition

People working in criminal justice rely on a number of tools to make their jobs easier. Professors keep their lectures fresh by incorporating the latest research published in scholarly journals. Practitioners search for "best practices" to help them achieve improved outcomes with their clients. But, whether you are a professor, a student, a practitioner, a journalist, or a novelist, access to the vocabulary of crime and justice is a must.

This updated and expanded edition is designed to serve as one of those tools. It is intended to be a comprehensive, though not exhaustive, guide to the vocabulary of contemporary criminal justice. If readers require a greater depth of information, they can consult reference works, such as encyclopedias of crime and justice. In some cases, they may want to consult specialty dictionaries, such as those focusing on law, forensic science, or biographies. There are also a number of websites that offer more extensive treatment of selected topics. However, for many readers this dictionary will meet their needs.

Since the last publication, a vast number of changes in society and in the field of criminal justice argued for a new edition of this dictionary. For one, the face of crime is changing. Terrorism is now such an integral part of the crime picture that a number of universities have created special terrorism centers to study its contours and effects. Unfortunately, mass shootings are almost a weekly occurrence, creating an ongoing debate over guns and violence. In recent years, assaults against the environment have received increased attention, resulting in a growing number of criminologists who study so-called green criminology.

These and other changes in society are also making criminology and criminal justice increasingly interdisciplinary. For decades, criminology was dominated by sociology, but that is no longer the case. Other perspectives on crime, such as the biological and psychological, once dismissed as reductionistic and invalid, now enjoy a resurgence of interest with new advances in science.

Another noteworthy development is that the divide between the academic and applied worlds is finally narrowing. There is a growing movement toward translational criminology, a criminology that strives to make itself more relevant to policy and practice, particularly by identifying evidence-based practices that reduce crime and improve the treatment of offenders.

Taken together, all these changes mean that there is a tremendous benefit for professionals to know the latest expanded terminology in criminal justice and related fields. *The Concise Dictionary of Crime and Justice* facilitates this exchange of vocabularies by offering a cross-section of terms from a wide variety of allied criminal justice professions.

How to Use This Dictionary

The most obvious way for the reader to use this dictionary is to look up unfamiliar terms. However, with the quick Internet access available to small phones, tablets, and computers, it is often easier to search using those devices.

I recommend the best way to use this dictionary is to simply browse through the contents. The undergraduate student in criminal justice might run across a term that provides the initial spark of an idea for a term paper. Academic criminologists, including those in graduate training, might make a connection between their areas of research and terms found in the dictionary, leading to new lines of inquiry. A specific example might be scholars who focus on wrongful convictions. If they encounter types of unfamiliar forensic equipment (e.g., a *mass spectrometer*), they might decide to research the extent that such devices could inadvertently contribute to the conviction of innocent persons. Likewise, a juvenile court judge might be pleasantly surprised to discover in this resource a program titled *Blueprints for Healthy Youth Development*, a potential source of evidence-based practices to implement in court.

This dictionary offers parallels for other professionals whose occupations also touch on crime and justice. The professional writer—whether nonfiction or fiction—might stir the muse by perusing the entries. Imagine that a writer who specializes in tales of serial murder discovers the term *anatomy murder* in this guide; this might lead to the speculation whether a person in modern times could commit a series of murders in order to supply medical schools with cadavers. A new novel plot is born.

In order to facilitate these connections, this dictionary is designed to help users find other related topics of interest. As a result, many of the entries suggest the user "see" or "compare with" other entries. These referrals help place each entry in a more meaningful context.

The primary purpose of this dictionary is to stimulate the reader's imagination. Recent developments in criminology and criminal justice have occurred in part because individuals thought outside of the box and no longer felt bound by the disciplinary blinders that prevent forward thinking. Likewise, the users of this dictionary may be able to synthesize information in the entries into new practices, new ideas for research and writing, and even new careers.

Acknowledgments

I would like to thank Jerry Westby at SAGE who inherited the project in its early stages and later recognized there was room in the marketplace for an updated and expanded revision. An author could not have a more supportive, patient, and understanding editor. Other folks at SAGE who have provided tremendous support include Laura Kirkhuff, Natalie Cannon, Kelly DeRosa, and Terra Schultz. Janet Ford, my copy editor, deserves a special thanks for devoting countless hours to carefully shaping both content and form. I am grateful for all these people have done to help make this book possible.

I must recognize the contributions of Rick Cooper in the genesis and development of this dictionary. Rick, a longtime friend and fellow writer, identified numerous entries for the first edition and gave it a thorough review. His friendship and support have always meant a lot to me.

Finally, I must thank my wife Jan. Knowing I have more project ideas than I can ever develop, she periodically prods me to bring at least some of them to completion so that I won't have any regrets later. I very much need and appreciate those nudges.

About the Author

Mark S. Davis is an independent scholar whose work focuses on crime and justice. His scholarship has appeared in journals, such as *Journal of Research on Adolescence*, *Social Psychiatry & Psychiatric Epidemiology*, and *Journal of Criminal Justice*, among other journals. Dr. Davis earned his Ph.D. in sociology from Ohio State.

abduction the unlawful taking of a person by force, fraud, or persuasion. Abduction is similar to *kidnapping* except that no demand for *ransom* is involved. An example of abduction is when a parent who does not have legal custody takes their child and hides the child's location from the custodial parent.

abet to encourage or assist an *offender* in the commission of a crime. See *aiding and abetting*.

abeyance the state of a criminal sentence when it is suspended. When legal consequences are held in abeyance, the convicted *offender* generally must abide by certain conditions.

abolitionism the movement to abolish the *death penalty* as a form of punishment. Abolitionism gained momentum in the late 20th century with the advent of *DNA testing* and its ability to exonerate the wrongfully convicted, including those awaiting execution. Abolitionists are particularly visible and vocal around the time of executions, often demonstrating outside a *prison* where an *execution* is about to take place. See *Bedau, Hugo Adam, Death Penalty Information Center, Innocence Project*.

aboriginal courts courts that hear and adjudicate cases involving *aboriginal crime*.

aboriginal crime crime committed by aboriginal or native persons, such as aboriginal tribes in Australia and New Zealand. See *indigenous crime*.

abortion the premature termination of a pregnancy. Criminal abortion is the act of intentionally producing a miscarriage or termination of pregnancy by any illegal means. Abortion became legal in the United States with the 1973 Supreme Court ruling in *Roe v. Wade*. Abortion remains a divisive issue, spawning periodic acts of violence by right-to-life extremists, including *premeditated murder* against abortionists and violence against abortion clinics.

abrasion collar a round hole with blackened margins made when a bullet pierces the skin. Abrasion collars tell investigators not only that the wound

A

was made by a firearm, but may also reveal the caliber, the distance between the shooter and victim, and other noteworthy forensic details. See *entrance wound, stellate, tattooing.*

ABSCAM   short for Arab Scam. ABSCAM was a scandal occurring from 1978 to 1980 where United States government officials were caught on tape accepting bribes from *Federal Bureau of Investigation* agents posing as representatives of Arab sheiks. The officials, which included members of the U.S. Congress, were convicted on a variety of charges. It was disconcerting to many that these public officials were so easily bribed. See *bribe, political crime.*

abscond   to secretly depart or flee a specific jurisdiction, especially one having legal control over an *offender.*

absolve   to release a person from any penalties, obligations, or consequences arising from a criminal act.

Abt Associates   a research and consulting firm based in Cambridge, Massachusetts which was founded by Dr. Clark Abt in 1965. Abt Associates uses research to address a variety of social problems and to manage projects related to criminal justice.

Abu Ghraib   a post-911 military correctional facility in Iraq where U.S. military personnel abused, humiliated, and photographed enemy prisoners of war. Several of the perpetrators involved were punished. See *human rights violations, khaki-collar crime.*

abuse   to injure physically, emotionally, or verbally. Abuse is also the result of such injury. See *child abuse, elder abuse, spousal abuse.*

abuse excuse   term used to describe the justification that an offender's past abuse is responsible for the criminal conduct in question. The abuse excuse is used by some offenders and their attorneys as a *mitigating circumstance* of the alleged offense. Critics argue that the abuse excuse is used to shirk *offender* responsibility for wrongdoing. See *neutralization techniques.*

abuse of a corpse   the criminal misuse of a dead body. This offense can take many forms, from the improper handling, storage, or molestation of the deceased by funeral home personnel to acts of *necrophilia* or *necrophagia* by serial killers and other offenders.

abusive home   a home where abuse occurs. The abuse can be physical, sexual, psychological, emotional, or verbal in nature; the victim of the abuse can be a person of any age. Abusive homes can pose a threat not only to the children who live there, but also to successive generations of offspring indirectly affected by the abuse. See *child abuse, elder abuse.*

Academy of Criminal Justice Sciences (ACJS)   an international membership organization that promotes professional and scholarly work in criminal justice. Established in 1963, the ACJS publishes two journals, *Justice Quarterly* and the *Journal of Criminal Justice Education*. ACJS holds its annual meetings in March of each year. The ACJS is headquartered in Greenbelt, Maryland.

Academy of Experimental Criminology (AEC) a scholarly body founded in 1998 to recognize selected scholars for their outstanding achievements in randomized controlled field experiments in criminology and criminal justice. The AEC is affiliated with the Division of Experimental Criminology of the *American Society of Criminology*.

Academy Group, Inc. a private forensic behavioral consulting firm composed of former members of the FBI's *Behavioral Analysis Unit* and other agencies who offer professional consultation and training services related to aberrant and violent behaviors. The Academy Group is headquartered in Manassas, Virginia. See *criminal investigative analysis*, *profiling*.

accessory a person who may not be directly involved in the commission of a crime, but assists the offender. See *accessory after the fact, accessory before the fact*, *complicity*.

accessory after the fact a person serving as *accessory* after the actual commission of the crime. Compare with *accessory before the fact*.

accessory before the fact a person serving as *accessory* prior to the actual commission of a crime. An example of an accessory before the fact is someone who lent assistance to an *offender* prior to the actual event without actually participating in the crime itself. Compare with *accessory after the fact*.

accidental killing the unintentional taking of another's life. See *involuntary manslaughter*.

accomplice a person who knowingly and willingly assists another person or persons in the commission or concealment of a criminal act. See *aiding and abetting*. Compare with *codefendant*.

accusation an *allegation* that a person has committed a crime. An accusation can be made without the initiation of formal charges.

accused a person or persons charged in a court of law with a crime, offense, wrongdoing, or fault. Compare with *defendant, person of interest, suspect*.

accuser one who brings an *accusation* against another. Often the accuser is the *victim*.

acquaintance rape the *rape* of an individual by someone known to that person. The majority of sexual assaults against girls and unmarried women are by someone they know. Compare with *date rape, intimate partner violence*.

acquit   to find a *defendant* not guilty by *jury*, *judge*, or panel of judges.

acquittal   the outcome of a criminal case when the *defendant* is set free from the charge of an offense by verdict of a *jury*, *judge*, sentence of a court, or other legal process.

actus reus   a Latin term that literally means a wrong deed. Actus reus is an act, which when combined with *intent*, constitutes a *crime*. See *mens rea*.

Adam Walsh Child Protection and Safety Act   a federal law enacted in 2006 which defines three tiers of sex offenders and requires them to comply with lifetime registration requirements; failure to do so is a felony. See *Walsh, Adam, Walsh, John, sex offender, sex offender registration*.

ad curiam   a Latin term meaning before a court or to a court.

Add Health   abbreviation for the *National Longitudinal Study of Adolescent Health*.

addiction   a dependence on drugs, alcohol, or a certain habit. Addiction is composed of physiological dependence and/or psychological dependence.

ADHD   abbreviation for *Attention Deficit Hyperactivity Disorder*.

adipocere   a fatty soap-like substance that can form when human or animal bodies decompose in moist, oxygen-deprived environments. Also referred to as grave wax and corpse fat.

adjudication   the process of disposing of a juvenile or criminal matter; the determination, decision, or sentence, especially without imputation of guilt. Compare with *disposition*.

adjudicatory hearing   a formal court hearing at which a youth's case in *juvenile court* is disposed. At the adjudicatory hearing, the youth generally is placed on *probation* or, in cases of violent or otherwise serious delinquency, is sentenced to confinement in a correctional institution for youths.

Administrative Office of the United States Courts   a federal office created in 1939 that supports the federal judiciary in the United States through a variety of services.

admissibility   the status of a statement or evidence that permits it to be allowed or conceded in criminal proceedings against an accused offender.

admissible   anything entitled or worthy of being introduced in court. See *testimony* and *evidence*.

admonish   to judicially warn or disapprove. Judges and magistrates sometimes admonish those before the court for behavior which is disrespectful or otherwise unacceptable.

Adonis, Joseph (1902–1971) a New York *mobster* considered by many to be instrumental in the evolution of American *organized crime*. See *Mafia*.

adult generally an individual who has reached his or her eighteenth birthday. With respect to criminal responsibility and consequences, adults are subject to harsher penalties than their juvenile counterparts. Compare with *juvenile*.

adult bookstore a business that sells *pornography* and sometimes sexual aids intended for use by adults. See *community standards*, *First Amendment*.

adulterated specimen a urine specimen submitted for *drug testing* which has been tampered with through the introduction of other substances. Compare with *dirty urine*.

adulteration the intentional or unintentional modification of *evidence* so as to render it useless for prosecution purposes. An example of adulteration includes the inadvertent contamination of evidence at crime scenes or problems with the subsequent handling of evidence.

adultery the act of a married person willfully engaging in sexual relations with someone other than the person's spouse. Although adultery was once punishable by death in many cultures and remains a crime in some jurisdictions, it has gained a certain degree of social acceptance.

adversarial justice the system of justice in which there are two opposing parties. Western criminal justice processes are based on an adversarial system where the prosecution and defense oppose one another.

adversary the opposing party in a legal dispute. Example: a *prosecutor* is an adversary to a *defense attorney*.

affective violence violence that is the result of highly charged emotions. An example of affective violence is where an individual, after a period of excessive drinking, reacts violently to an insult from another.

affidavit a voluntary sworn written statement of facts made especially under oath or on affirmation before an authorized magistrate or officer.

affirm the act of validating an earlier decision or ruling. When an appellate court affirms a decision by a lower court, the earlier decision stands.

affirmative defense an acceptable rebuttal to a legal proscription (an imposed restraint or restriction) against a certain type of behavior. For example, in many jurisdictions, an affirmative defense to the charge of *carrying a concealed weapon* is that the accused is a business person who regularly transports large sums of money, and thus needs to carry a weapon for self-protection. An affirmative defense does not keep a person from being charged with a crime; however, if used successfully, it can result in dismissal of the charges.

affray   a fight between two or more persons in a public place causing a disturbance to others. See *disorderly conduct*.

AFIS   abbreviation for *Automated Fingerprint Identification System*.

aftercare   a period of supervised control of a releasee from a juvenile correctional facility. Aftercare is an opportunity to ensure compliance with special conditions, such as treatment or *restitution* and to receive counseling and other necessary social services. Compare with *parole*.

aftercare worker   a worker who provides *aftercare* services.

agent provocateur   a spy, or one employed to associate with suspected persons, often to infiltrate an organization for the purpose of collecting intelligence and to pretend sympathy with their aims in order to incite them to some incriminating action. See *espionage*.

aggravated assault   a *physical assault* where serious bodily injury occurs or where a weapon capable of inflicting such injury is used.

aggravating circumstance   a circumstance surrounding a crime that serves to increase its seriousness and the severity of the penalty. An example of an aggravating circumstance is the use of a firearm in the commission of a robbery. Compare with *mitigating circumstance*.

aggression   hostile action that is potentially injurious to another. Aggression can be proactive or reactive.

agricultural crime   a *crime* unique to agricultural regions and settings. Examples of agricultural crimes include the theft of livestock and illegal dumping. See *rural criminology*.

aiding and abetting   to willingly and deliberately assist another in the commission of a crime. See *accomplice*.

aid panel   primarily used in New South Wales, a group consisting of a *police officer*, a *solicitor*, community members, and young persons who work with the court to identify opportunities for youthful offenders.

air piracy   the illegal commandeering of an aircraft by force or threat of force. Also, informally known as skyjacking. An example of air piracy is the September 11, 2001 takeover of commercial jets that were later crashed into the World Trade Center towers and the Pentagon. See *Attack on America, hijacking, piracy*.

AK-47   originally called Automat Kalishnikov, a Soviet-made, fully automatic *assault rifle* and one of the most widely used weapons in the world.

AK-47s are frequently mentioned in discussions of banning assault weapons. An AK-47 was used in the *mass murder* of school children in Stockton, CA in 1989. See *automatic weapon*.

alarm   a device that gives off a sound or signal calling attention to some event or condition. It is often intended to alert the occupants of a building to some violation or intrusion. See *burglar alarm, car alarm*.

Alcatraz   an island and site of the former federal *penitentiary* in San Francisco Bay. Alcatraz was made infamous by some of its inmates, including *Al Capone*, George "Machine Gun" Kelly, Alvin Karpis of Ma Barker's gang, and Robert "The Birdman" Stroud. Alcatraz was used as a prison for difficult-to-manage offenders from 1934 to 1963. It was known as a prison from which escape was nearly impossible due to the strong currents and sharks in the Bay. From 1969 to 1971, Alcatraz was occupied by Native Americans who were trying to reclaim Indian land, as well as bring attention to the plight of the American Indian. Since 1973, Alcatraz has been a popular tourist attraction. See *federal prison*.

alcohol   ethyl alcohol, the intoxicating substance found in beer, wine, liquor and other spirits. Alcohol has been linked in many ways to criminal behavior, including *victim-precipitated crime*.

alcohol and other drugs (AOD)   an umbrella term used to represent the vast array of substances of abuse as well as the general field of those who specialize in the prevention and control of *substance abuse*, including alcoholism.

Alexis, Aaron (1979–2013)   the civilian contractor who perpetrated the *Washington Navy Yard Shooting*. Alexis was killed by police.

Alford plea   a plea, named for the court decision, *North Carolina v. Alford*, where the defendant neither admits guilt nor claims innocence, but admits that the prosecution can likely prove the charge. Compare with *no contest*.

alias   a pseudonym assumed by a criminal for the purpose of avoiding detection or capture. Used especially in legal proceedings to connect the different names of anyone who has gone by or been known by two or more names.

alien   one who is not a citizen or legal resident of a country. See *criminal alien, illegal alien, undocumented alien*.

allegation   a written or verbal statement, often before a court, claiming that someone has committed a crime.

allege   to make a formal claim that someone has committed a crime.

Alliance of NGOs on Crime Prevention and Criminal Justice   a group of non-governmental organizations interested in criminal justice issues that consult with the United Nations.

allocution   the right in common law of an individual to offer written or oral statements before a court. The right of allocution often is used as the legal basis for introducing *victim impact statements* and *private presentence investigation reports* in court.

All Points Bulletin (APB)   a law enforcement announcement broadcast to authorities throughout a wide jurisdiction to be aware of an offender, missing person, or other law enforcement emergency. Compare with *Be on the Lookout (BOLO)*.

alpha   a statistical concept in criminological research that represents the probability that a person will re-offend. Alpha has played a significant role in research on the deterrent effects of various legal punishments. See *deterrence*.

al Qaeda   a Middle Eastern terrorist organization once headed by *Osama bin Laden*. It is believed that al Qaeda is responsible for a number of terrorist attacks against U.S. facilities, including the *Attack on America*. See *international terrorism, terrorism*.

al Shabaab   a jihadist and *terrorist* organization based in Somalia that is affiliated with *al-Qaeda*. See *jihad*.

altercation   a sometimes loud angry dispute between two or more people often escalating to a physical attack.

alternative dispute resolution (ADR)   a set of methods as well as a movement to find ways of settling disputes outside traditional judicial processes. Alternative dispute resolution includes the prevention of disputes as well as their peaceful resolution.

amateur detective   a lay person who engages in the activities of a *detective*. Some mystery novels feature amateur detectives. See *crime novel, mystery*.

AMBER alert   an alert issued to the public by law enforcement that a child is missing and presumed kidnapped. Named for nine-year-old Amber Hagerman who was kidnapped and murdered in 1996, AMBER alerts have resulted in the successful rescue of children and the apprehension of their abductors.

ambush   a surprise attack perpetrated by a person or persons unseen by the victim prior to the attack.

American Academy of Forensic Sciences (AAFS)   a membership organization of physicians, criminalists, toxicologists, attorneys, document examiners, and others interested in forensic science. Headquartered in Colorado Springs, Colorado, the AAFS publishes the *Journal of Forensic Sciences* and holds its annual meetings every February.

American Bar Association (ABA)   an American professional organization of more than 400,000 lawyers. Headquartered in Chicago, the ABA offers a variety of services, including law school accreditation, continuing legal education, and other programs designed to assist lawyers, judges, and other members of the legal profession. The ABA has standing committees on *gun violence* and *substance abuse.*

American Bar Foundation (ABF)   the funding and research arm of the *American Bar Association.* The American Bar Foundation has sponsored and conducted a number of influential *socio-legal studies.* Staffed by full-time employees as well as a number of visiting fellows, the ABF maintains close working relationships and shares resources with Northwestern University and the University of Chicago. The ABF is headquartered in Chicago, Illinois.

American Board of Criminalistics (ABC)   a forensic science board composed of regional and national organizations which establishes standards for certification of professionals in the field of *criminalistics.*

American Board of Forensic Odontology (ABFO)   a professional organization whose objective is to establish and raise standards related to *forensic odontology.* The ABFO offers board certification to qualified professionals.

American Civil Liberties Union (ACLU)   a national organization dedicated to preserving the rights of individuals. Criminal justice issues which have involved the ACLU include *police brutality, racial profiling,* and the state of indigent defense. The ACLU frequently takes legal action against organizations whose activities threaten to endanger basic rights.

American Correctional Association (ACA)   a national organization of more than 20,000 correctional practitioners founded in 1870. The ACA, which holds an annual Congress of Corrections and a winter conference is headquartered in Alexandria, Virginia. Its publications include *Corrections Compendium, Corrections Today,* and *Correctional Health Today.*

American Jail Association (AJA)   a national organization that supports people who operate and work in jails. The AJA, which is headquartered in Hagerstown, Maryland, publishes the magazine *American Jails.*

American Law Institute (ALI)   an organization of lawyers and legal scholars whose purpose is to promote the clarification and simplification of the law and its adaptation to social needs. Established in 1923, the ALI has an elected membership of 3,000 lawyers, judges, and law professors. Among the accomplishment of the ALI is the development of the *Model Penal Code.* The ALI is headquartered in Philadelphia, Pennsylvania.

American Probation and Parole Association (APPA)   an international organization whose members work or have an interest in *parole, probation,*

A

and other forms of *community corrections.* The APPA is headquartered in Lexington, Kentucky.

American Society for Industrial Security (ASIS)   an international organization of security professionals. Headquartered in Alexandria, Virginia, ASIS works to increase the effectiveness of security through educational programs and materials. It publishes the magazine *Security Management.*

American Society of Criminology (ASC)   an organization formed in 1941 that supports and promotes criminology as a distinct professional field. The American Society of Criminology in concerned with the entire spectrum of the process of criminal justice, as well as scholarly inquiry leading to new theory and knowledge. Each November, the ASC holds its annual meetings for its members. The ASC publishes *Criminology: An Interdisciplinary Journal, Criminology & Public Policy,* and a newsletter, *The Criminologist* and is headquartered in Columbus, Ohio.

American Society of Victimology (ASV)   a national organization for those who identify with the field of criminology. The ASV, which holds periodic symposia and colloquia, promotes evidence-based practice and education about *victimology.*

America's Most Wanted (AMW)   a television program that ran from 1988 to 2012 whose mandate was to highlight wanted felons in hopes that the viewing audience could assist law enforcement in identification and apprehension. AMW, which was hosted by *John Walsh,* claimed that the program resulted in more than 1,000 captures.

amicus curiae   Latin term meaning friend of the court. Also, a person with an interest in a matter before a court who files a brief in support of one of the parties.

amicus curiae brief   a brief filed by an *amicus curiae* on behalf of one of the parties.

amido black   a protein stain used by forensic scientists to enhance patterns and details in blood. See *bloodstain pattern analysis.*

ammunition   a cartridge consisting of the projectile and its casing and propellant used in a firearm. Also referred to as ammo. See *ballistics.*

amnesia   partial or total loss of memory, often arising as a result of trauma to the brain. Those accused of crimes have been known to claim amnesia as a defense; in some of these cases, the accused feigns amnesia.

amnesty   the granting of a *pardon* by government to an individual or group.

Amnesty International an independent, non-political organization whose mission it is to protect *human rights* around the world. See *Universal Declaration of Human Rights*.

amphetamines any of several central nervous system stimulants. Active ingredients can include amphetamine, dextroamphetamine, or methamphetamine. Although some amphetamines have proper medical uses, such as medically supervised weight loss, they are also the drug of choice among many drug abusers. Street names include speed and uppers.

Anastasia, Albert (1902–1957) a powerful *organized crime* figure who once served as head of *Murder, Inc.* and later boss of the Gambino crime family. He was shot and killed in a barber shop in Manhattan. See *Mafia*.

anatomy murder murder specifically for the purpose of supplying cadavers for medical research and teaching.

angel of death a type of *serial killer*, often a nurse, medical technician, or other health professional who kills patients in their care. Some angels of death believe that they are relieving their victims of suffering by taking their lives. Infamous angels of death include *Donald Harvey* and *Michael Swango*.

anger management efforts or programs aimed at curbing negative, unhealthy human emotions and their expression in aggression and violence.

Angola the site of the Louisiana State Penitentiary in Angola, Louisiana. In the past Angola made news because of its deplorable conditions. It is also the prison where executions take place in Louisiana.

animal abuse the gross neglect or cruelty to animals.

animus the Latin term for intention or motivation.

anomie a state of normlessness or lawlessness thought to be due in part to homogeneity in the population. See *anomie theory, Durkheim, Emile, strain theory*.

anomie theory a sociological theory, first articulated by French sociologist *Emile Durkheim* and later expanded on by sociologist *Robert K. Merton,* that posits that deviance occurs when there is an unequal emphasis in society on the ends people are expected to achieve and the means available to achieve them. Anomie theory is *structural* in that it attributes pathology, such as crime, to social forces rather than to individual pathologies. Later theoretical restatements include Richard Cloward and Lloyd Ohlin's book *Delinquency and Opportunity*. Compare with *General Strain Theory, strain theory*. See *Ohlin, Lloyd*.

anoxia the deprivation of oxygen to the body. Significant brain damage or death can result if the period of anoxia is prolonged. Cerebral anoxia is a common cause of death for a person who commits suicide by hanging.

antemortem   a Latin term meaning occurring before death. This term is frequently used to describe wounds or other conditions that occurred before a person died. Compare with *postmortem.*

Anthony, Casey (1986–present)   a young Florida woman who was accused of killing her two-year-old daughter Caylee. The case resulted in a sensational trial where Casey Anthony was acquitted of murder, but convicted on other charges. See *filicide.*

Anthrax case   a criminal case in 2001 where anthrax spores were sent to selected individuals by mail, resulting in five deaths and the infection of 17 others. *Bruce Ivins* was considered by the FBI to be responsible for these crimes. See *bioterrorism.*

anthropometry   the measurement of the human body and its constituent parts for the purpose of classification, identification, and analysis. See *forensic anthropology.*

Anti-Defamation League (ADL)   a national organization committed to exposing and fighting anti-Semitism and other forms of hatred and bigotry, including white supremacism and *Holocaust* denial. The ADL, founded in 1913, maintains regional offices around the United States and international offices. See *hate crime.*

antidote   any substance that counteracts the effects of a *poison* or other toxic substance. See *poisoning.*

antiquities theft   the theft of valuable relics. Compare with *art theft.*

antisocial behavior   behavior which does not conform to ordinary standards of decency or acceptability.

antisocial personality disorder   a personality *disorder* characterized by superficial charm, lack of empathy, and a disregard for the rights of others. Antisocial personality disorder is described in the *Diagnostic and Statistical Manual of Mental Disorders.* Those suffering from antisocial personality disorder resist efforts at treatment. Compare with *psychopathy, sociopath.*

antitrust laws   federal and state laws designed to prevent *price-fixing* and monopoly control.

Anttila, Inkeri (1916–2013)   a Finnish law professor and *criminologist* who was heavily involved in the reform of the Finnish Penal Code and in making *criminology* relevant to criminal justice policy.

AOD   abbreviation for *alcohol and other drugs (AOD).*

Apalachin, NY   the site of a 1957 meeting of *organized crime* figures from around the country. Law enforcement authorities raided the estate where the meeting was being held, detaining the attendees and recording their license plate numbers. Apalachin served as evidence that there indeed was a *Mafia* and that its tentacles reached across the entire United States.

apartheid   an official policy of racial segregation. See *racial justice*.

APB   abbreviation for *All Points Bulletin*.

appeal   a post-disposition legal process in a criminal or civil case in which one of the parties formally argues in writing to higher courts that substantial mistakes were made at the lower court level and requests a rehearing.

appeal bond   a special bond that permits a convicted offender to remain free pending the outcome of an appeal. In the absence of an appeal bond, the offender begins serving the sentence imposed by the lower court.

appeals court   a court higher than the court of original jurisdiction of a case where appeals are made and where appeals are decided. Also referred to as appellate court. There are twelve federal courts of appeal each covering a group of states called a "circuit." See *supreme court*.

appearance   the presence of a criminal *defendant* in *court*.

appearance bond   a *bond* to guarantee a defendant's appearance at future court hearings.

appellant   the party that appeals to a higher court for a review of a lower court's decision. See *appeals court*. Compare with *appellee*.

appellee   the party in an appeal that argues the correctness of the lower court's decision. See *appeals court*. Compare with *appellant*.

applied criminology   the application of criminological theory and research to criminal justice policy and practice. Compare with *public criminology*, *theoretical criminology*, *translational criminology*.

appointed counsel   a private attorney appointed by the court to represent an indigent client. Appointed counsel often is used in jurisdictions not having a public defender. There are those who believe the modest fees attorneys receive in such cases serve as a disincentive to the preparation of a strong defense. Compare with *public defender*.

apprehend   to capture, arrest, and take into custody a *suspect* in a crime.

arbitration   a dispute resolution where the two parties work to settle their disagreement through an arbiter and agree to abide by the final decision.

arch   a segment of a human *fingerprint*. Compare with *loop, whorl*.

argot   a special vocabulary or slang unique to a group of people. Groups having their own argot include prison inmates, street gangs, and those involved in the selling and using of drugs.

ARIMA   acronym for auto regressive integrated moving averages, a statistical technique for forecasting trends in criminal justice data. ARIMA, which has been used to predict future crime rates and correctional populations, incorporates the lagged effects of variables on subsequent events. See *prison population forecast*.

Armed Career Criminal Act   a federal law in the United States that provides severe sentences for offenders convicted more than twice of violent felonies or serious drug offenses.

armed robbery   a robbery where a weapon is used or feigned by the *offender*. Armed robbery is one of the more serious felonies. It is also known as aggravated robbery, in which the use of a weapon, most often a firearm, is the *aggravating circumstance*.

aroma scan   a process employed in *forensic science* for sensing and analyzing vapors, gases, and related aromas emanating from corpses, arson scenes, meth labs, and other sources.

Arpaio, Joe (1932–present)   the *sheriff* of Maricopa County, Arizona whose controversial methods of dealing with offenders have included housing them in tents and putting them to work on chain gangs. See *chain gang*.

arraignment   the first appearance in the court of jurisdiction. During an arraignment, the *defendant* typically enters an initial plea, the court ensures representation by counsel, and bond is set or continued. Compare with *initial appearance*.

arrest   the taking into custody of a person by the police or other legal authorities with the intention of pressing criminal charges.

arrest clearance   the official removal by police of an active case following the arrest of a suspect. Because arrest clearances are considered a measure of police effectiveness, there is an incentive to clear arrests by any means, including the admission by offenders who may have had nothing to do with the crime in question.

arrestee   one who has been arrested. Compare with *defendant, detainee, suspect*.

Arrestee Drug Abuse Monitoring (ADAM)   a national effort sponsored by the U.S. Department of Justice to routinely collect data on drug use by those who are arrested and jailed. ADAM data show which drugs are being used by arrestees, and which drugs fall in and out of popularity. Trained data collectors interview arrestees and collect specimens at the various ADAM sites. ADAM includes males and females and both adults and juveniles. See *Drug Use Forecasting (DUF)*.

arrest order   a written order issued by a *parole officer* or *probation officer* to arrest a parolee or probationer for a new offense or a *technical violation*. Compare with *warrant*.

arrest practices   the various means by which law enforcement officers effect arrests. Arrest practices can be controversial if they are thought to be abusive, discriminatory, or otherwise unfair or inappropriate. See *police brutality, racial profiling*.

arrest record   a complete list of arrests for a given individual. Compare with *conviction record*.

arrest statistics   data on the number and characteristics of people arrested for crimes. Compare with *offense statistics*.

arrest warrant   a warrant issued by a judge giving law enforcement the authority to take a specific individual into custody.

arsenic   a poisonous white powder used to commit murder and suicide. Arsenic is found in rat poison and certain herbicides. See *Swango, Michael*.

arson   the malicious and unlawful burning of a building or other property. Arson includes the destruction of one's own property for *fraudulent insurance claims*. See *fire setting, serial arson*.

arson accelerant   a substance like gasoline or lighter fluid used to accelerate the combustion of a fire. See *arson*.

arsonist   an individual who commits *arson*. Compare with *fire setter*. See *serial arsonist*.

artificial fibers   fibers that come from certain types of clothing, carpet, rope or other synthetic materials that may be of forensic interest. See *hairs and fibers*.

art theft   the theft of fine art, such as paintings or sculptures. Compare with *antiquities theft*.

Aryan Brotherhood   a white supremacist organization formed in San Quentin prison in 1967 to protect its members from Blacks and Hispanics inside the institution. Drug trafficking is a major source of income for the group. The Aryan Brotherhood has been responsible for numerous violent crimes, including murder.

Aryan Nations   a group of neo-Nazi extremists that believe in the superiority of the white races and is dedicated to their preservation. See *Aryan Brotherhood, Ku Klux Klan*.

A

Asian Criminological Society   a professional organization whose mission is to promote the study of criminology and criminal justice across the Asian continent. The Asian Criminological Society is headquartered in Macau, China.

asphyxia   the deprivation of oxygen leading to unconsciousness, injury, or death. One way to die from asphyxia is by *suffocation*. See *anoxia, burking*.

asphyxiate   to suffocate or cause unconsciousness as a result of interference of the exchange of oxygen and carbon dioxide in the body. See *burking, suffocation*.

assailant   one who commits or is suspected of committing *assault*.

assassin   one who plans, attempts, or carries out an *assassination*.

assassination   the *premeditated murder* of a prominent person by surprise attack, often for political or religious reasons. Infamous 20th-century assassinations include those of President John F. Kennedy, the Rev. Dr. Martin Luther King, Jr., and former Beatle John Lennon. See *assassin*.

assault   the unlawful threat or touching of another person with intent to do bodily harm. Assault is often confused with *battery*. Also referred to as simple assault. Compare with *aggravated assault, felonious assault*.

assault rifle   an *assault weapon* of *rifle* length.

assault weapon   an automatic or semiautomatic firearm, generally with a large capacity magazine, designed for firing a high volume of ammunition within a short period of time. An example of an assault weapon is the *AK-47*. There have been many legislative efforts to define assault weapons and to control their manufacture, distribution, ownership and possession.

assembly line justice   a term used to convey justice processes so routinized that they compromise true justice. The notion of assembly line justice is reinforced by practices, such as the *plea bargain,* which in many courts occurs in the vast majority of cases.

asset-focused approach   the identification and use of an offender's positive assets, such as family and community support, as opposed to focusing on risks and deficits. Compare with *risk-focused approach*.

asset forfeiture   the legal requirement that certain accused or convicted offenders surrender real or other property believed to be obtained from their illegal activities. Asset forfeiture gained popularity in the 1980s with law enforcement agencies involved in the investigation of *drug trafficking*. Compare with *asset seizure*.

asset seizure   the taking by the government of money, property, or other items gained illegally through criminal activity.

assisted suicide   the taking of one's own life with the help of another, often a physician or other medical professional. Assisted suicide is illegal in most jurisdictions. See *Kevorkian, Jack*.

Association of Chinese Criminology and Criminal Justice   a scholarly and professional organization whose purpose is to promote research and education on Chinese criminology and criminal justice.

asylum   historically, a shelter, such as a church or temple, that offered protection from arrest or persecution. Also, the subject of a book *The Discovery of the Asylum*, by historian David Rothman where he traced the history of prisons and mental hospitals.

atavism   a characteristic in an offender thought to be related to an earlier, more primitive form of being. For example, because a prominent brow was characteristic of Cro-Magnon man, a similar feature on an offender might prompt some to believe that criminality is linked to less evolved forms of Homo sapiens. Nineteenth century Italian criminal anthropologist *Cesare Lombroso* advanced the notion that certain types of offenders were throwbacks to an earlier form of evolutionary being, and therefore could be identified by certain physical characteristics. See *Kretschmer, Ernst*.

at-risk youth   a youth who by personal, family, community, or cultural characteristics is deemed vulnerable to engaging in deviant or delinquent behavior, but who has not become involved in the *juvenile justice* system.

atrocity crimes   particularly repugnant crimes, such as *genocide* and *war crimes*. See *human rights violations, International Criminal Court*.

attachment   one of the four elements of Travis Hirschi's control theory of delinquency. Hirschi hypothesized that youths attached to their families and conventional values stand a greater chance of being insulated from delinquency. See also *belief, commitment, control theory, involvement*.

Attack on America   term used by politicians, the media, and others to describe the September 11, 2001, terrorist attacks on the World Trade Center and the Pentagon as well as related terrorist plots against the United States. See *al Qaeda, international terrorism, bin Laden, Osama, terrorism*.

attempted crime   a crime which has not been completed. Attempted crimes generally are punished slightly less severely than the corresponding *completed crime*.

attendance centre   a place in the United Kingdom where youthful offenders regularly report as imposed by a court.

attention center   same as *detention center*.

A

attention deficit hyperactivity disorder (ADHD)   a childhood disorder whose symptoms include difficulty staying focused, controlling behavior, and over-activity.

Attica   the site of the Attica Correctional Facility in Attica, New York, that became a buzzword for prison reform after a bloody *prison riot* in September of 1971. Inmates who were protesting crowded living conditions and possible racial overtones in inconsistent sentences and parole decisions took over cell blocks for four days. The uprising ended when police stormed the facility and retook control with 10 correctional officers and civilian employees and 33 inmates dying and over 80 wounded in the process.

attorney-client privilege   the long-standing tradition of confidentiality that exists between attorneys and those they represent. Attorney-client privilege theoretically prevents the disclosure of information a client divulges to an attorney.

attorney general   the chief legal officer of the federal or state government. Attorneys general are appointed at the federal level, but are often elected at the state level. Their responsibilities include representing the government in legal proceedings.

Auburn system   a system of prison discipline in the 19th century characterized by strict policies for the prisoners: segregation in cells at night, walking in lock step, maintaining silence, congregate work in shops during day, and dining seated back to back while communicating with hand signals. The Auburn system was first employed at the Western State Penitentiary in Pennsylvania.

audit trail   a series of financial documents that when linked together can support fiscal responsibility and correlatively uncover embezzlement, fraud, or other types of financial misconduct or crime. Audit trails are important in the investigation of *organized crime* and *white-collar crime*, whose perpetrators often go to great lengths to hide their illegally gotten assets through tangled webs of complex financial transactions.

Augustus, John (1785–1859)   a Boston cobbler generally regarded as the father of *probation*. Beginning in 1841, Augustus supervised alcoholics and youths under an agreement with the local court.

Australian and New Zealand Society of Criminology (ANZSOC)   the scholarly and professional organization to promote research and training in the field of criminology in Australia and New Zealand.

Australian Bureau of Statistics   the governmental agency in Australia which promotes the collection of high-quality statistical data for decision making, including data related to crime and justice.

Australian Institute of Criminology (AIC) the Australian government agency responsible for conducting and supporting criminological research. The AIC issues numerous publications of interest to researchers, policy-makers, and practitioners.

autoerotic fatality a fatality that results from the practice of *autoeroticism*. Practitioners of autoeroticism often use elaborate props and restraints to carry out their fantasies, most commonly with some type of hanging or neck compression to reduce the flow of oxygen into the body. Sexual gratification is achieved as the individual approaches unconsciousness. A self-rescue safety mechanism generally is incorporated into the practice, however in the case of fatalities, unconsciousness sets in before the escape method can be used. These deaths are often mistakenly deemed suicides or homicides by investigators unfamiliar with this phenomenon. In other cases, the true circumstances surrounding these deaths are kept from relatives to spare them emotional pain or public embarrassment.

autoeroticism self-arousal and sexual satisfaction by means of fantasy or genital stimulation. See *autoerotic fatality*.

Automated Fingerprint Identification System (AFIS) a system that permits the electronic collection, storage, retrieval, and comparison of human fingerprints. These systems read, match, and store fingerprints. While this does not entirely eliminate the need for manual examination of fingerprints, it speeds up the process by reducing possible matches. See *fingerprint, LiveScan*.

Automated Property System a system designed to permit law enforcement agencies to make use of information related to pawn brokers and second-hand dealers. See *fence, receiving stolen property*.

automatic weapon a firearm that fires continuously as long as the trigger is depressed, or until the ammunition is expended. Compare with *semi-automatic weapon*.

autopsy a postmortem examination of the internal and external parts of a body to determine or confirm the *cause of death* and *manner of death*. Same as *postmortem*.

autosadism the infliction of pain on one's self for sexual gratification. Compare with *sadism, self-injurious behavior*.

auto theft the theft of an unoccupied automobile, truck, or other similar vehicle. Auto theft is one of the major offense types which comprise the *Uniform Crime Reports* Crime Index. See *carjacking, unauthorized use of a motor vehicle*.

A

aversion therapy   a form of psychodynamic intervention where the patient is subjected to unpleasant sensations (e.g., electric shock) in conjunction with an image to be extinguished. For example, pedophiles have been treated with aversion therapy by viewing pictures of children while receiving a mild electric shock, theoretically conditioning them to thereafter associate sex with children with pain. Also referred to as aversive conditioning.

avulsion   the tearing away of a body part or tissue as a result of trauma or a surgical procedure. The examination of avulsions by medical personnel or forensic investigators often can point to the kind of weapon used in a violent crime. See *entrance wound, exit wound.*

baby boom a noticeable increase in the birthrate within a relatively short period of time. Usually, this term refers to the post-World War II population explosion of children fathered by returning veterans. Baby booms increase the population, which causes increases in crime when the "baby boomers" reach offending ages. The same phenomenon causes problems in correctional populations as confined offenders age and suffer health and other age-related problems. This bulge in the population has also been descriptively referred to as a "pig in a python." Compare with *echo boom*.

Baby Face Nelson see *Gillis, Lester.*

background check the result of investigating the criminal, financial, military, or other facets of an individual's personal history associated with the purchase of a firearm.

background investigation an investigation undertaken to determine the employment, financial, or criminal history of an individual, often when they are being considered for employment.

backlog in a criminal court, the cases awaiting disposition. A number of solutions have been sought for the problem of case backlogs, including *diversion*. See *caseload reduction*, *court delay, speedy trial*.

backup additional support from other law enforcement officers in order to promote safety and effect an arrest.

bad check a check written with insufficient funds in the bank account to cover it. The writing of a bad check can be unintentional or intentional, the latter sometimes is defined as *passing bad checks*. See *check kiting*.

badgering the harassment of a *witness* giving testimony in a court of law. This tactic is often employed by attorneys in order to make strong impressions on the jury or to confuse a witness into making contradictory statements.

bad seed hypothesis a theory that certain human beings are born with the capacity to engage in criminal behavior. Once dismissed, this hypothesis has more credence with the advent of genetic evidence that suggests the

B

inheritability of criminal traits. See *biocriminology, biosocial criminology, constitutional theory*.

bad time   days added to the sentence of an imprisoned offender for poor conduct. Bad time became popular with the advent of a fixed *determinate sentence* and the more punitive philosophy of *retribution* of the 1970s and 1980s. Also known as bad time credit. Compare with *good time credit*.

bagging of hands   the tying of paper bags around the hands of homicide victims in order to preserve evidence. If the victim struggled with the attacker, bits of skin and blood may be found under the former's fingernails. Paper bags are used because plastic bags cut off the air flow, increasing the rate of decomposition and potentially degrading any *trace evidence* that might be present. See *fingernail scraping*s.

bail   money, property, or other security offered in exchange for the release from custody of an arrested person and to guarantee their appearance at trial. Bail is forfeited if the accused does not appear in court. See *bond, Manhattan Bail Project, pretrial release*.

bail agent   a person who finds and takes into custody those who have skipped out on bail bonds. While the tactics of bail agents are often surrounded by controversy, higher courts have upheld the constitutionality of their work. Some states, most notably California, mandate standardized training for those who want to serve as bail agents.

bail bond recovery agent   see *bail agent*.

bail bondsman   a person who makes a living ensuring the *appearance* of criminal defendants in return for a fee, often a percentage of the bail bond set by the court. Bail bondsmen are supposed to forfeit the bond to the court if the defendant fails to appear, but in practice this transaction seldom happens. See *surety*.

Bailey, F. Lee (1933–present)   a famous criminal defense attorney who represented a number of notorious individuals, including *Sam Sheppard* and *O. J. Simpson*.

bailiff   an officer of the court whose job entails controlling access to the *judge* or *magistrate*, keeping the judge's calendar and maintaining order in the courtroom. Generally, bailiffs are appointed by the judges for whom they work.

bail reform   any movement to make the practice of bail more equitable or effective. Studies have shown that bail practices of the past discriminated against certain segments of the population, including minorities and the poor. See *Manhattan Bail Project, pretrial release*.

B

Bail Reform Act of 1984 a federal act that permits judges to consider the potential dangerousness of offenders in non-capital cases when appropriate bail is being set. This act replaced the Bail Reform Act of 1964.

bait and switch a deceptive practice used by some retailers to persuade a customer to buy a different, often more expensive item rather than the advertised item that drew the customer to the store. This practice is illegal if the advertised item was never actually available. See *false advertising*.

balanced and restorative justice (BARJ) an approach to justice which strives to promote community protection while ensuring offender accountability and *reconciliation* with victims. See *reintegrative shaming, restorative justice*.

ballistic knife a spring-loaded or gas-powered tube from which a bladelike projectile is propelled. Ballistic knives are regulated by federal law in the United States. Compare with *switchblade*.

ballistics the scientific study of projectiles, either still in the bore of the firearm or after the weapon is fired. Microscopic examination of the markings on a spent bullet can determine the type of weapon fired. Ballistics is also a term used to describe the labs or units that conduct studies of projectiles, their trajectories, and the specific firearms used.

banditry a crime, often *robbery* and sometimes *murder,* committed by outlaw bands.

banishment a punishment where the *offender* is forced to leave his or her original homeland, sometimes to a specified alternative location for a specified period of time. Banishment was popular in the 18th and 19th centuries in Europe. It has a more limited value in modern times because many nations now refuse to accept convicted offenders from other countries. Compare with *transportation*.

Banopticon a term coined by Didier Bigo to describe the use of *profiling* to decide which individuals in society should be placed under surveillance. Compare with *panopticon*.

bar any authority, court, or tribunal that renders judgment or makes a final evaluation. Also, all the judges, prosecutors, attorneys, and other legal practitioners who participate in the local justice system.

Barbie, Klaus (1913–1991) a former Nazi Gestapo commander during World War II who was charged with committing atrocities against the Jews. Barbie was known as "the butcher of Lyons." See *concentration camp, Holocaust, war crimes*.

barbiturates a group of sedative drugs derived from barbituric acid. Physiologic effects include decreased blood pressure, respiration, and heart rate.

B.A.R.J. or BARJ   abbreviation for *balanced and restorative justice*.

B

Barker-Karpis gang   an American criminal gang in the early 1930s involved in numerous bank robberies and kidnapping. See *Barker, Ma*.

Barker, Ma (1873–1935)   in the early 1930s, a woman who with her sons committed a range of robberies and murders. Barker was killed in a shoot-out with law enforcement officers. See *Barker-Karpis gang*.

Barnes, Harry Elmer (1889–1968)   a historian of the early to mid-20th century who chronicled the history of confinement, torture, and execution in his book, *The Story of Punishment*. Barnes was criticized later in life for his denial of the *Holocaust*.

barrister   in England, a counselor who has been admitted to the *bar* and thus can advocate for clients. Compare with *counselor, solicitor*.

barroom violence   violence that occurs in or around a bar, tavern, or other establishment where *alcohol* is sold for consumption on the premises. Barroom violence, studied as early as the 19th century by Belgian statistician *Adolphe Quetelet*, is fueled by the intoxication of the parties and facilitated by the carrying and use of weapons. See *victim-precipitated crime*.

Barrow, Clyde (1909–1934)   a notorious robber of the 1930s who was responsible for the deaths of several people, including law enforcement officers. Barrow and *Bonnie Parker* were ambushed and killed by officers led by retired Texas Ranger Frank Hamer. See *Texas Rangers*.

baton   a long stick fashioned of wood or synthetic material designed to permit law enforcement officers to bring suspects under control without the use of *lethal force*. See *billy club, nightstick, PR-24, truncheon*.

Batson challenge   a challenge made by either the prosecution or defense during *jury selection* alleging that a *peremptory challenge* was used to exclude a *juror* due to race, ethnicity, or gender.

battered child syndrome   a psychological condition brought on by a pattern of behavior that develops in children who are subjected to abuse that occurs regularly over time, finally resulting in the conclusion that violence against the parents or caregivers is the only way to end the abuse.

battered wife syndrome   a psychological condition brought on by a pattern of behavior that develops in women who are subjected to a form of spousal abuse that occurs regularly over time, finally resulting in the conclusion that violence against the husband is the only way to end the abuse. See *intimate partner violence*.

batterer   one who batters or a *domestic violence* perpetrator.

battery   intentional physical contact with another person, without that person's consent. Lay persons often confuse battery with *assault*.

bawdy house   a house of *prostitution*. Also referred to as *brothel*, house of ill repute, whorehouse.

beat   refers to the geographical area routinely patrolled by a law enforcement officer.

Beccaria, Cesare Bonesa, la Marchese di (1738–1794)   an Italian nobleman and intellectual of the late 18th and early 19th centuries who wrote a famous essay titled *On Crimes and Punishment*. In his book, Beccaria spoke out against a number of unjust practices of the day, including *torture*, secret accusations, the abuse of power by judges, and inconsistencies in sentencing offenders. Beccaria's book was a success in Europe, appearing at a time when intellectuals were questioning a number of social institutions of the day. See *classical school of criminology*.

Becker, Howard S. (1928–present)   an American sociologist best known for his contributions to the sociology of deviance, and in particular the *labeling perspective*. Becker spent most of his career at Northwestern University. His books include the sociological text *The Outsiders: Studies in the Sociology of Deviance*.

Bedau, Hugo Adam (1926–2012)   a Harvard-trained philosopher who became famous for his writings and presentations in opposition of *capital punishment*. A founder of the *National Coalition to Abolish the Death Penalty*, Bedau spent most of his career at Tufts University. His books include *The Death Penalty in America* and *Killing as Punishment*. See *death penalty*.

bed wetting   the involuntary enuresis of children or adolescents. Bed wetting is considered one of the three early indicators of future homicidal behavior. The other two indicators are *fire setting* and *cruelty to animals*.

Behavioral Analysis Unit (BAU)   the unit of the *Federal Bureau of Investigation* that uses in-depth crime scene analyses, criminal *profiling*, and other techniques to solve homicides and other violent and serial crimes. Its mission includes the development and provision of training, research, and consultation programs that are grounded in the social and behavioral sciences. The BAU applies expertise developed studying serial killers to other crimes, such as public corruption and white-collar crimes. The BAU is headquartered at the FBI National Academy in Quantico, Virginia and is part of the *National Center for the Analysis of Violent Crime (NCAVC)*. See *criminal investigative analysis*.

behavioral case linkage   connecting criminal cases through the analysis of behavioral cues left by the *offender*. See *linkage blindness*.

behavioral evidence analysis   see *criminal investigative analysis.*

behavior modification   the process of altering the maladaptive behavior of individuals through the use of classical conditioning and operant conditioning. See *differential association reinforcement theory.*

belief   one of four elements of Travis Hirschi's control theory of delinquency. See *attachment, commitment, control theory, involvement.*

belly chain   a steel chain designed to pass through the belt loops of an individual in custody to which handcuffs are attached. This prohibits the individual so constrained from using the arms as weapons. See *handcuffs, leg irons, manacles, shackles.*

Beltway Sniper   name given by the media for the person or persons sought in the sniper shootings in and around Washington, D.C. in 2002 that left 10 people dead and three critically injured. *John Muhammad* and *Lee Malvo* were arrested and convicted for the crimes.

bench   where a judge sits and conducts business in a courtroom.

bench trial   a trial in which the verdict is rendered by the presiding *judge*. In bench trials, the defendants must waive their Constitutional right to *trial by jury*. Bench trials can be presided over by a single judge or in some cases by a three-judge panel. Compare with *jury trial.*

bench warrant   a *warrant* issued by a judge.

Bentham, Jeremy (1748–1832)   a jurist and philosopher of the late 18th and early 19th centuries. Bentham is best known for his writings which advocated for the abolition of the *death penalty* and the imposition of punishments commensurate to the seriousness of the crime. He cast man as hedonistic, but who could freely choose among various courses of action, including those defined as criminal. See *classical school of criminology.*

Be on the Lookout (BOLO)   a law enforcement advisory requesting that other agencies be aware that a wanted suspect might be in their jurisdictions. Compare with *All Points Bulletin, AMBER alert.*

Berkowitz, David (1953–present)   a *serial killer* who was responsible for the deaths of six men and women and the wounding of seven others in New York in 1976 to 1977. These crimes were referred to as the Son of Sam murders. After he was arrested, Berkowitz claimed that he was taking orders from a possessed dog. He was sentenced to serve six consecutive life sentences. See *Son of Sam laws.*

Bertillon, Alphonse (1853–1914)   a French police records clerk who developed the *Bertillon classification system.*

Bertillon classification system a now archaic system of using various anthropometric measurements to identify and categorize criminals, named for its developer *Alphonse Bertillon*. Consistent with prevailing criminological thought of the day, the Bertillon system consisted of measurements of the head and body to classify offenders. It was eventually discredited due to its inability to ensure uniform measurement, and by the advent of fingerprinting which proved more reliable. See *criminal anthropology*, *Cesare Lombroso, positive school of criminology*.

bestiality sexual intercourse between a human being and an animal. An individual charged with bestiality would likely be charged with *cruelty to animals*. Also referred to as *zoophilia*.

Bestie di Satana a *satanic cult* in Italy responsible for three grisly ritual murders in 1998 and 2004.

best practice a criminal justice practice or program that consistently yields better outcomes than other programs. See *Blueprints for Healthy Youth Development*. Compare with *evidence-based practice*.

betting wagering or *gambling* on any of a number of activities, including sporting events, horse and dog races, card and dice games, or other games and activities governed by skill and/or chance.

Bhopal disaster a large gas leak at the Union Carbide pesticide plant in Bhopal, India which killed more than 2,000 people and sickened or disabled numerous others. It is considered one of the worst industrial disasters in world history. The incident resulted in fines and prison terms for several Union Carbide employees. See *corporate crime*.

Bianchi, Kenneth (1951–present) a *serial killer* who, with *Angelo Buono, Jr.*, was responsible for the murders of numerous female victims. Because of the location of their Los Angeles area murders, Bianchi and Buono became known as the Hillside Stranglers. Bianchi is serving life imprisonment in the state of Washington for murders committed there.

bias crime crime motivated by a bias against a particular class of individuals based on race, ethnicity, age, sexual identity, or disability. Compare with *hate crime*.

bifurcated process a two-part trial. An example of a bifurcated process is a capital trial where the first part of the process centers on a finding of guilty or not guilty, the second part generally has the jury deciding between *life imprisonment* and the *death penalty*.

bigamy the criminal offense of being legally married to more than one spouse at the same time. Despite its illegality, bigamy continues to be practiced by some persons of the Mormon faith and others. There are instances of men who maintain two families at the same time.

Big Brothers Big Sisters   an organization whose mission is to match mentors to young people, particularly those from disadvantaged backgrounds who are at risk of becoming involved in delinquency.

big house   slang term for a *prison.*

bill of attainder   any act of a legislative body condemning a person or group of persons guilty of a crime and assessing a punishment without a formal trial. Bills of attainder are prohibited by the Constitution of the United States.

bill of indictment   see *indictment.*

bill of information   a charging document occasionally used by prosecutors when a felony case is not taken to a *grand jury.* Bills of information are sometimes used to charge those who are not going to contest the charges. Compare with *indictment.* See *prosecuting attorney.*

bill of particulars   a document citing specific allegations against a person.

Bill of Rights   the first ten amendments of the Constitution of the United States. It is the Bill of Rights that restricts the role of federal government by guaranteeing a number of individual freedoms, such as protections for those suspected or accused of committing crimes, and granting the right to bear arms.

billy club   a rod-like weapon, often fashioned from hardwood, used by law enforcement officers to bring arrestees under control. Compare with *night stick, PR-24, truncheon.*

binge drinking   the practice by college students of drinking large amounts of alcoholic beverages in a short period of time. Binge drinking has been defined as four or more alcoholic drinks within one hour. While binge drinking is not in and of itself a crime, those who binge drink may engage in criminal conduct, including *assault, vandalism,* and *indecent exposure.*

bin Laden, Osama (1957–2011)   a terrorist leader considered responsible for a number of terrorist acts, including the World Trade Center and Pentagon attacks in the United States in 2001. Osama bin Laden, born into a wealthy Saudi Arabian family, renounced his Saudi citizenship, was disowned by his family and thereafter committed himself to fighting Western enemies of Islam. Osama bin Laden was shot and killed by U.S. Navy Seals in 2011, and his remains were buried at sea. See *al Qaeda, Attack on America.*

biocriminology   the branch of *criminology* that conducts research on the role of biological factors in the etiology of criminal behavior. See *constitutional theory, twin studies.*

biological criminology see *biocriminology*.

biological determinism the belief that certain biological factors are responsible for the genesis of criminal behavior. See *biocriminology*.

biological evidence evidence that derives from the human body, such as blood or tissue. It is biological evidence that permits officials to employ *DNA testing* to link offenders to specific crimes. Even in cases where bodies are severely burned, biological evidence can sometimes be obtained from bone marrow or the pulp of teeth.

biological fluids fluids from the human body, including blood, bile, and urine that can be analyzed for forensic purposes. Biological fluids are a subset of *biological material*.

biological material material coming from a human, including blood, urine, semen, saliva, bile, tissue, hair, and nails.

biometric sensor devices used to detect unique physical traits for identification and security. Biometric sensors are used to scan eyes, fingerprints, and *biological material*.

biosocial criminology a branch of *criminology* that includes both biological and environmental factors in trying to explain criminal behavior. Such factors can include inherited characteristics and neuropsychological deficits.

bioterrorism the use of biological material, such as anthrax or botulism, to perpetrate *terrorism*. Long a potential threat, bioterrorism became more real in the wake of the *Attack on America*. See *Anthrax case, Ivins, Bruce*.

birching a form of punishment where an individual is struck with a birch rod. Compare with *caning, flailing*.

Birmingham church bombing the bombing of the 16th Street Baptist Church in Birmingham, Alabama on September 15, 1963 that killed four young African American girls. The motivations for the bombing were *racist*. See *civil rights*.

birth cohort a group of individuals who have their year of birth in common. In a well-known longitudinal study, criminologists *Thorsten Sellin* and *Marvin E. Wolfgang* tracked the criminal careers of a birth cohort of males born in 1945 to determine how many went on to become involved in serious delinquency.

birth order the order in which a person is born in relationship to siblings. Birth order has been studied by criminologists as one of many possible factors involved in the *etiology* of criminality.

bite-mark identification identification of a suspect through the impression of tooth marks left behind in the skin of the victim or in other materials, such as chewing gum or food. Often used in conjunction with saliva washings of the bite mark area on the skin, which can yield blood groups on serological examination. Bite-mark identification has come under attack due to the lack of scientific consensus on this specialized field. See *forensic odontology, junk science.*

bivariate analysis   the analysis of two variables. An example is the analysis of the relationship between socioeconomic status and delinquency. Compare with *multivariate analysis.*

Black criminology   a study of criminology informed by the social reality of being Black in society. Black criminology moves beyond the analysis of race to include the historical treatment of Blacks and the involvement of Black criminologists.

Black Dahlia   name given to homicide victim Elizabeth Short whose mutilated body was found in Los Angeles in 1947. The case was never solved.

Black Hand   an extortionist group active in the United States in the early part of the 20th century. The Black Hand has its origins in Sicily, Italy. Also referred to as *Mano Nera.*

black hat   a computer *hacker* whose motive is maliciousness or material gain.

blackjack   a small concealable club-like weapon often constructed of leather and filled with lead.

blackmail   the illegal act of demanding payment or other benefit from someone, under the threat of physical or other harm if payment is withheld. Compare with *extortion.*

black market   the illegal trafficking in merchandise that is either in scarce supply or whose manufacture or distribution is heavily regulated.

Black Panther Party   a group founded in 1966 by Bobby Seale and Huey P. Newton. The Black Panthers were known for advocating the end of racial discrimination and economic oppression of Blacks and were also involved in food giveaways and free health clinics. The Black Panther Party was investigated by the *Federal Bureau of Investigation* for alleged criminal activities. See *Cointelpro.*

Black Talon   a discontinued special bullet, outlawed in many jurisdictions, that causes maximum damage to human tissue due to sharp points that extend on impact. Similar bullets are still available on the market. Compare with *cop-killer bullets.*

black tar   a type of *heroin* named for its tar-like color and consistency. Black tar comes from Mexico.

blast the concussive result of an explosion.

blasting agent an explosive that cannot be detonated by blasting caps when unconfined. An example is ammonium nitrate mixed with diesel fuel, which is the type that was used in the *Oklahoma City Bombing*.

blended sentencing a mixture of juvenile and adult sentences for juvenile offenders. In blended sentencing, a juvenile may be sentenced to serve time in a juvenile institution, but may later be transferred to an adult institution on reaching age 18. Blended sentencing became popular in the 1990s during the movement toward tougher measures for juveniles.

block watch a group of neighbors in housing subdivisions or other cluster-type developments who join forces for the purpose of watching for and reporting suspicious people or activities in their neighborhood. Police departments often assist in forming and advising block watch groups. Signs are posted prominently on the streets in the neighborhood to warn criminals that their activities might be observed by local block watch members. See *crime prevention, Zimmerman, George*.

blood alcohol concentration see *blood alcohol content*.

blood alcohol content (BAC) the percentage of ethanol in an individual's blood. BAC can be detected by a *Breathalyzer* or by the analysis of a blood sample. See *drunk driving*.

Bloods a notorious Los Angeles *street gang* that has spread across the United States. Compare with *Crips, Latin Kings*.

blood spatter the distribution of blood as the result of an impact or gravity. See *bloodstain pattern analysis*.

blood spatter analysis the analysis of the amount and distribution of blood at a crime scene. See *bloodstain pattern analysis, castoff*.

bloodstain the residue of blood. Bloodstains reveal numerous clues to crime investigators, including not only blood type, but also the position of the victim and other details of the crime. See *bloodstain pattern analysis*.

bloodstain pattern analysis (BPA) a forensic science specialty involving the in-depth analysis of bloodstains in order to determine details about a crime.

blotter a book or other media where entries or occurrences are recorded of police arrests and activities pending transfer to permanent record books.

bludgeon to strike with a heavy club or club-like object. Also, the club or object used in the attack.

bluebeard a term given to a man who serially marries and then murders his wives, based on a fictional character in a story by Charles Perrault.

B

Blueprints for Healthy Youth Development   a list of evidence-based prevention and intervention programs for youths identified by the Center for the Study and Prevention of Violence at the University of Colorado. See *evidence-based practice*.

blunt force trauma   a severe impact to the head or body as the result of a blunt instrument or tools, such as baseball bats, lumber, or large rocks. This type of trauma results in internal injuries. Compare with *sharp force trauma*.

bobby   a police officer in London, U.K., named for *Sir Robert Peel*.

body armor   bulletproof vests and other clothing designed to stop the penetration or lessen the impact of bullets. See *bulletproof vest*, *Kevlar*.

body camera   small cameras worn by law enforcement officers in order to record interactions with citizens. Sometimes body cameras are used to photograph crime scenes.

body farm   a parcel of land where donated human remains are permitted to decay for the purpose of studying decomposition, insect activity, and other processes of scientific and forensic interest. See *cadaver*, *forensic entomology*.

body snatcher   one who steals corpses. See *anatomy murder*, *burking*.

body type theory   a dated criminological theory which posited that the body types of individuals were related to various temperaments, including those related to criminality. Body type theory is most closely associated with psychologist *William Sheldon*.

Boesky, Ivan (1937–present)   an infamous Wall Street broker of the 1980s found guilty of *insider trading*. Boesky used confidential stock information to illegally gain millions of dollars. In return for leniency, he agreed to help authorities convict *Michael Milken* and others involved in the same scheme. See *white-collar crime*.

bogus   phony; not genuine. Bogus is often used to describe counterfeit money or credentials, such as passports or driver's licenses.

boiler room fraud   a *scam* where the perpetrators work out of rented offices known as boiler rooms. Using banks of telephones, they solicit donations for non-existent charities or to interest potential investors in fraudulent companies. See *telemarketing fraud*.

Boko Haram   a militant Islamist *terrorist* organization based in northeast Nigeria dedicated to preventing the adulteration of Islamic practices through Westernization. Compare with *al Qaeda*, *Islamic State in Iraq and Syria*.

BOLO   an acronym used by law enforcement for *Be on the Lookout.*

bomb squad   a unit of a law enforcement agency specially trained to safely handle, move, disarm, and detonate bombs and explosives.

B

bomb technician   a police officer or civilian who is trained to identify, diffuse, and dispose of bombs.

Bonanno, Joseph (1905–2002)   a Sicilian-born *Mafia* boss and head of the Bonanno *crime family.*

bond   a written guarantee of performance. See *bail.*

bondage   sadomasochistic activities generally involving bindings and restraints, hoods, gags, and blindfolds. Bondage can be practiced with a single partner, in groups, or alone. Sometimes bondage is taken too far, resulting in serious injury or death. See *autoeroticism, sadomasochism.*

bond agent   see *bail bondsman.*

bonding   see *bond.*

bondsman   see *bail bondsman.*

bookie   a bookmaker; a person who takes illegal bets. See *gambling.*

booking   the recording in official police records of the facts pertaining to an individual's arrest, including identifying information and the specific charges that were filed. It is common for a suspect to be fingerprinted as part of the booking process.

bookmaking   the illegal taking of bets. See *gambling.*

booster   a thief, particularly a type of professional shoplifter.

booster girdle   an elastic band or similar device worn around the waist or other part of the body of a shoplifter for the purpose of holding stolen goods tightly against the body in order to conceal them. See *shoplifting.*

boot camp   a correctional facility for youthful or first-time offenders designed to operate in a similar manner as a military boot camp. Boot camps use structure, discipline, physical training, and verbal intimidation by drill instructors to break down and rebuild offenders. Research has shown that boot camps, while effective in helping offenders in the short term, have not been as successful in obtaining long-term changes in behavior.

bootlegger   a person who engages in the illegal manufacture, sale, or transportation of *alcohol.* The term is derived from the practice of *smuggling* alcohol by concealing it in an actual bootleg. Bootlegging was prevalent during the U.S. experience with *Prohibition.* See *Bureau of Alcohol, Tobacco, Firearms, and Explosives (ATF), Volstead Act.*

Borden, Lizzie (1860–1927) a Fall River, Massachusetts spinster who is alleged to have viciously murdered her affluent father and stepmother with a hatchet in 1892. Lizzie Borden was tried and found not guilty by a *jury*. She was still considered guilty by many. The case was never solved.

borderline personality disorder a *personality disorder* characterized by instability of relationships and impulsive behavior. Those with borderline personality disorder have an unstable self-image and may engage in *self-injurious behavior,* including *suicide*. Borderline personality disorder is a common diagnosis among confined female offenders.

Borstal a residential facility in Britain where youthful offenders learn a trade, such as masonry or carpentry, receive an education, and participate in counseling. Youthful offenders would spend anywhere from six months to two years at the Borstal. Dating back to the early 1900s, the Borstal takes its name from Borstal, England where the first such facility was opened.

boss the head of an organized *crime family*.

Boston Marathon bombings the detonation of two pressure cooker bombs during the Boston Marathon on April 15, 2013 that resulted in three deaths and numerous serious injuries. These crimes have been attributed to brothers *Dzhokhar Tsarnaev* and *Tamerlan Tsarnaev*.

Boston Strangler the nickname given by the media to a rapist and murderer who terrorized Boston in the early 1960s. The murders were eventually attributed to *Albert DeSalvo*.

botanical material fragments of plants that can be used in solving crimes. Botanical material may be found on suspect's shoes, in the tire treads of vehicles or other obscure places. The analysis of botanical material, for example, can lead investigators to the original crime scene in cases where the offender transported the body after the crime. See *forensic botany*.

botched execution an execution of a condemned person where there is a malfunction of the method of execution, resulting in unanticipated pain or discomfort, or a protracted death. Examples include drugs used for *lethal injection* that fail to bring about a quick and painless death of the condemned.

bounty hunter a privately employed agent who tracks down and apprehends wanted and fleeing felons in return for reward money. Bounty hunters have broad powers under the Constitution of the United States, and, as such, are not well-regulated. Consequently, they have been at the center of much controversy over some of the questionable tactics used in locating and apprehending felons. Compare with *bail agent*.

Brady Bill a piece of federal legislation, named after former White House press secretary James Brady, who was shot and disabled during the 1981 *assassination* attempt on President Ronald Reagan's life. The Brady Bill was designed to reduce the opportunities for criminals to obtain handguns by requiring purchasers of handguns to undergo a criminal background check. See *gun control, Handgun Control, Inc., National Instant Criminal Background Check System.*

Brady Center to Prevent Gun Violence a nonprofit organization dedicated to reducing injuries and death due to firearms through public health and safety programs. See *Brady Bill.*

Brady Gang a gang of robbers and murderers composed of Al Brady and others who were eventually shot and killed by FBI agents in Bangor, Maine in 1937.

Brady material from *Brady v. Maryland,* information or evidence that could be exculpatory to a criminal defendant. See *exculpatory evidence.*

Brady Rule a rule that requires prosecuting attorneys to disclose *exculpatory evidence* to the defendant's attorney. It takes its name from *Brady v. Maryland.* Failure of a prosecutor to do so can result in the suppression of the evidence. See *disclosure.*

Branch Davidians a group who lived at a complex near Waco, Texas and followed the teachings of David Koresh. Koresh and a number of the Branch Davidians died in a shootout with federal authorities in 1993. This controversial incident allegedly served as motivation for the *Oklahoma City bombing.* See *Bureau of Alcohol, Tobacco, Firearms & Explosives.*

branding an archaic form of punishment inflicted by applying a hot iron brand to the flesh of an individual, creating a serious burn and leaving a permanent, disfiguring scar.

brass knuckles a handheld device, fashioned from brass rings or other cast metals, used to inflict greater harm when hitting a person with the fist.

Brawner test the legal test of mental illness that asserts that the person is not responsible for the crime if at the time of the crime, the conduct was the result of mental disease or defect such that either the individual is incapable of appreciating the wrongfulness of his conduct or he cannot conform his conduct to what is required by the law. See *insanity defense.*

break-in the forced entry into a building for the purpose of committing a crime. See *breaking and entering.*

breaking and entering the felonious entering of a dwelling or business for the purpose of committing a theft offense. Compare with *burglary, criminal trespass.*

Breathalyzer a machine that measures the percentage of alcohol in a person's bloodstream. Breathalyzers are used by law enforcement officers to test the *blood alcohol content* of *driving while intoxicated* suspects. The accused breathes air through a plastic tube which is analyzed by an alcohol simulator. A digital readout displays the results of the measurement. See *drunk driving, ignition interlock.*

bribe   money given illegally to another in return for a favor or consideration.

bribery   the act of giving, offering, or accepting a *bribe.*

Bridewell house   a prison or house of correction in England.

brief   a legal document on behalf of a client that asserts a position, poses a legal question, or makes a request.

brigand   dated term to describe an offender who lives by plunder and usually travels in a band.

Brink's robbery   a robbery of a Brink's armored car in New York in 1981 during which $1.6 million was stolen. Two police officers and a security guard were killed during the robbery.

British Crime Survey   former name of the *Crime Survey for England and Wales.*

British Society of Criminology   the professional membership organization of criminologists in the United Kingdom. The British Society of Criminology holds an annual conference and publishes the journal *Criminology & Criminal Justice.* It maintains its headquarters in London, England.

broken windows   see *broken windows theory*

broken windows probation   a form of probation supervision grounded in the *broken windows theory* that helped spawn *community policing.* Broken windows probation connects probation more closely with the problems and needs of the communities where probationers live and are supervised, focusing on programming that works.

broken windows theory   a theory based on an influential and often cited 1982 article in *The Atlantic Monthly* by criminologist George Kelling and political scientist *James Q. Wilson.* Kelling and Wilson asserted that outward manifestations of community decay, such as broken windows and unmown vacant lots lead to more serious decay of urban neighborhoods, creating fear in residents and opening the door to violence and other criminal behaviors.

brothel   a house or other structure where prostitution takes place. Also *bawdy house*, house of prostitution, whorehouse.

Brussel, James (?–1982)   a psychiatrist and *criminologist* who gained recognition as an early *profiler* in highly-publicized criminal cases in the mid-20th century. He is the author of *Casebook of a Crime Psychiatrist*. See *criminal investigative analysis.*

BTK Killer   nickname given to the *serial killer* who murdered 10 people in and around Wichita, Kansas from 1974 to 1991. BTK stands for stands for "Bind, Torture, Kill," which was his infamous signature. *Dennis Rader* was eventually caught and convicted for the crimes.

buggery   an unnatural sex act between humans or between a human and an animal. Compare with *bestiality, sodomy, zoophilia.*

Bulger, James "Whitey" (1929–present)   an *organized crime* figure and federal informant in Boston, Massachusetts who was responsible for numerous murders and other crimes. Bulger eluded law enforcement for 16 years until his capture in 2011. He is serving a life sentence in a federal *penitentiary.*

bullet jacket   a sheath, usually made of copper, that encases a lead bullet for the purpose of reducing lead fouling of the inside of the gun barrel. Bullet jackets can be analyzed to determine characteristics of the bullet and the gun from which it was fired.

bulletproof vest   a vest that is impenetrable by bullets. Worn by many police officers, but also utilized by criminals. With the application of *Kevlar*, bulletproof vests became much more effective in providing protection for the wearer. This, in turn, was compromised by the use of Teflon-coated bullets which can pass through such vests. Consequently, some bulletproof vests incorporate impenetrable steel plates. Compare with *body armor.* See *cop-killer bullets.*

bullet track   the path made by a bullet entering and passing through the body. Examination of bullet tracks can yield useful information about the position and angle of the weapon when fired and the relative positions of the perpetrator and victim. See *ballistics, entrance wound, exit wound, trajectory.*

bull pen   a holding area in a jail or lockup for those awaiting arraignment or transfer to another facility.

bullying   threatening or intimidating behavior, sometimes accompanied by physical violence, by some youths toward others, especially those smaller, weaker, or otherwise perceived to be vulnerable. Bullying has been the target of specially tailored prevention and intervention programs. See *cyberbullying, Olweus Bullying Prevention Program.*

bullying prevention   initiatives designed to preventing *bullying* and bullying-related behaviors. See *Olweus Bullying Prevention Program.*

Bundy, Theodore "Ted" (1946–1989)   a prolific *serial killer* of the 1970s. Bundy abducted and murdered numerous girls and young women in the states of Washington, Utah, and Colorado. The former law student eventually was apprehended after he murdered two young women in a sorority house at Florida State University. He was convicted of these crimes, along with the rape and murder of a 12-year-old girl, and he died in Florida's *electric chair* in 1989.

Buono, Jr., Angelo (1934–2002) a *serial killer* who, with his cousin, *Kenneth Bianchi*, kidnapped, raped, and murdered 10 people in California in the late 1970s. Buono died of a heart attack in prison. See *Hillside Strangler*.

burden of proof the obligation on the part of the government to prove that an accused person is guilty of committing a crime. In the United States, such proof in a criminal case must be beyond a *reasonable doubt*, a standard higher than that required in civil proceedings.

Bureau of Alcohol, Tobacco, Firearms, and Explosives (ATF) an organizational unit of the U.S. Department of the Treasury responsible for the enforcement of federal laws pertaining to the manufacture and distribution of alcoholic beverages, tobacco products, and firearms. Their responsibilities encompass the investigation of bombings, such as that of the World Trade Center in 1993. The ATF drew much criticism for their role in storming the *Branch Davidians* complex in Waco, Texas where a number of men, women, and children died.

Bureau of Justice Assistance (BJA) a subdivision of the *Office of Justice Programs*, U.S. Department of Justice. BJA is responsible for administering formula and discretionary grant programs designed to prevent and control violent and other crimes, and to improve the administration of criminal justice. Among the grant programs administered by BJA is the Edward Byrne Memorial State and Local Law Enforcement Assistance Program.

Bureau of Justice Statistics (BJS) a subdivision of the *Office of Justice Programs (OJP)* within the U.S. Department of Justice that promotes the collection and analysis of crime data in the states and territories. BJS accomplishes this in part by awarding funds to state units known as *statistical analysis centers*. BJS has been instrumental in promoting the *National Incident Based Reporting System (NIBRS)*.

burglar a person who commits *burglary*. See *cat burglar*.

burglar alarm an electronic device designed to alert residents that a *break-in* has occurred. Burglar alarms use a variety of technologies, most recently cellular telephones to notify residents and authorities of an intrusion.

burglary the unlawful entry into an occupied dwelling for the purpose of committing a theft offense. This term is often confused with *robbery*. Compare with *breaking and entering*, *criminal trespass*.

burking murder by asphyxia and smothering in such a way as to disguise the manner of death by leaving the victim unmarked. Some cases of Sudden Infant Death Syndrome (SIDS) are suspected of actually being instances of burking. Burking is named for William Burke who along with William Hare killed people and sold their bodies to a doctor.

bushfire arson a form of *arson* in Australia where the bush is intentionally set on fire.

business crime crime committed in the course of transacting business. Examples of business crimes include *white-collar crime* and *corporate crime*.

bust an *arrest*.

butt the grip end of a *handgun* or stock end of a *long gun* opposite the *muzzle* end. Compare with *muzzle*.

bystander an individual in close proximity to a crime or accident; a chance spectator who can possibly affect the outcome of an incident by their presence. Compare with *innocent bystander.*

B

Cabrini-Green a former high-rise public housing complex in Chicago, Illinois known for its poor living conditions and gang-related violence.

cache a store of drugs, weapons, or other illegal goods.

cadaver a human corpse, especially one to be used for dissection and instruction in medical schools. See *anatomy murder*.

cadaver dog a dog specially trained to detect the odor of human decomposition. Cadaver dogs are used to locate human remains in the investigation of crimes and natural disasters. Some critics regard these alleged abilities as *junk science*.

cadaveric spasm the physiological reaction that can occur soon after death when a single muscle group becomes stiff and rigid.

CALEA abbreviation for the *Commission on Accreditation of Law Enforcement Agencies.*

caliber the inside diameter of a rifle or handgun barrel, or the diameter of a bullet or projectile. For example, the bore of a .22 rifle is 22/100 of an inch in diameter. Compare with *gauge*.

Cali cartel an infamous illegal drug organization based in Cali, Colombia. At one time it was thought the Cali cartel controlled the vast majority of the world's cocaine distribution. Its influence extended to government officials, including the military, the police, and other organizations and individuals. See *cartel, Medellin cartel, trafficking.*

California Criminalistics Institute a unit of California state government that offers training in *forensic science*.

California Youth Authority the state agency in California responsible for the confinement of youths remanded by county juvenile courts. The California Youth Authority distinguished itself in the latter part of the 20th century by participating in innovative programming, including the use of *I-level* classification system.

call for service a call received by a law enforcement agency indicating that someone needs police assistance. Calls for service often are employed by researchers and policy analysts as an alternative to crime statistics, such as those gathered for the *Uniform Crime Reports*. Plotting calls for service on a map can aid in the identification of *hot spots*.

call girl   a *prostitute* who uses the telephone to schedule appointments with clients. Call girls generally are accorded higher status than street prostitutes, a difference reflected in their income and lifestyle. See *Fleiss, Heidi, prostitution*.

Cambridge-Somerville Youth Study   a research study of a delinquency prevention project that began in 1939. The Cambridge-Somerville Youth Study was unique at the time because it carefully examined a large number of youths over a long period of time. Data collected on these youths included intelligence, personality characteristics, school progress, and neighborhood characteristics. See *longitudinal study*.

Cambridge Study in Delinquency Development   a *longitudinal study* undertaken by University of Cambridge criminologists where comprehensive social and family data were gathered on youths in order to test hypotheses about the correlates of delinquent behavior.

cameras in the courtroom   the use of video cameras to record court proceedings, most often trials. Cameras in the courtroom are controversial because they can disrupt the proceedings and possibly taint the viewing public's perception of events. Such controversies pit the freedom of the press as expressed in the *First Amendment* against the right to a fair trial as guaranteed in the *Sixth Amendment*. See *Court TV.*

Camorra   a criminal organization in Italy with a long history and extensive political reach. Compare with *Mafia*. See *organized crime*.

camouflage-collar crime   crime related to fish and wildlife. See *poaching, wildlife crime*.

Campaign for an Effective Crime Policy (CECP)   an initiative of criminal justice professionals and policy makers to move toward a "less politicized, more informed debate" on criminal justice policy. The CECP questions unnecessarily long prison sentences, punishment without treatment for substances abusers, the easy access to guns, and the elevation of highly publicized crimes to policy importance.

Campbell Collaboration   an initiative to produce systematic reviews of the effects of various social interventions, including those related to crime and delinquency. See *evidence-based practice, systematic review*. Compare with *Cochrane Collaboration*.

campus crime   crime occurring in or around a college or university campus. While campus crime encompasses all forms of crimes, it has come to mean those types of crime common to campus settings, such as *date rape*, bicycle theft, and offenses associated with *binge drinking*.

campus unrest a term used to convey a variety of disturbances occurring on or around college and university campuses. Examples of campus unrest include demonstrations against war, environmental policy, or other issues. See *civil disobedience*.

Canadian Criminal Justice Association (CCJA) a Canadian membership organization whose mission includes increasing greater understanding of criminal justice issues and improving the criminal justice system. The CCJA, which publishes the *Canadian Journal of Criminology and Criminal Justice*, is headquartered in Ottawa, Ontario, Canada.

caning the practice of punishing offenders by beating them with split cane shafts. Caning is particularly painful because as the cane is withdrawn, the splits close, pulling flesh with them. One infamous case of caning was that of an American citizen Michael Fay, whose 1994 caning in Singapore generated worldwide attention and pleas for leniency. See *corporal punishment, flogging*.

cannabis sativa the scientific term for *marijuana*.

cannibalism the act of one human being eating the flesh of another human. The most notorious practitioners include *Jeffrey Dahmer*, a *serial killer* who kept body parts of his victims in his refrigerator, and the fictional Hannibal Lector from the book, *The Silence of The Lambs*. In most cultures cannibalism is considered an offense *malum in se*. Compare with *necrophagia*.

canvass the process used by law enforcement officers to gather information about a crime or other incident under investigation by asking questions of witnesses, nearby residents, and other persons in close proximity to a *crime scene*.

capable guardian according to *routine activities approach*, an individual or organization that can prevent a motivated *offender* from committing a crime.

capacity the ability of a firearm's magazine to hold a specific number of cartridges. Some jurisdictions limit the capacities of magazines by law in an effort to limit the weapon's *lethality*. In corrections, the number of detainees or inmates a facility can, should, or does hold. See *design capacity, rated capacity*.

capias from Latin, a legal document issued by a *judge* or *magistrate* ordering the arrest of an individual.

capital crime a crime for which death is a potential penalty. See *capital punishment, death penalty*.

C

Capital Jury Project   a program of research designed to collect data on how juries exercise *discretion* in their decisions in *death penalty* cases. Based at the University at Albany, State University of New York, the Capital Jury Project is funded by the *National Science Foundation*. See *capital punishment, jury.*

capital offense   same as *capital crime.*

capital punishment   punishment that results in the death of the convicted *offender*. Capital punishment has become more controversial with exoneration of wrongfully convicted individuals through the use of *DNA testing*. See *death penalty.*

capo di tutti capi   Italian for *boss of all bosses*. See *boss, Mafia, organized crime.*

Capone, Al(phonse) (1899–1947)   a notorious *organized crime* figure of the 1920s and 1930s. Capone rose through the ranks of the Chicago mob to become one of the most powerful bosses and bootleggers in the United States. He eventually was convicted of federal *income tax evasion* and was sentenced to *prison*, serving some time in *Alcatraz*. After his release from prison, he died of syphilis. See *bootlegger, Prohibition.*

caporegime   often abbreviated as capo, literally means in Italian the head of a regiment or a lieutenant in an *organized crime* family. Compare with *boss.*

Capote, Truman (1924–1984)   an American writer whose book, *In Cold Blood*, chronicled the brutal 1959 slayings of the Clutter family in Kansas and the conviction and execution of the two killers.

car alarm   a horn or other device that makes a loud sound when the vehicle is tampered with.

carbine   a light short barreled *rifle*. An example is the M-1 carbine, a .30 caliber version of the longer, heavier Garand rifle used by U.S. forces in World War II.

car bomb   an explosive device placed in, under, or near a motor vehicle for the purpose of killing or injuring those in or around it. Car bombs are most closely associated with organized crime and terrorists and can have devastating effects not only to occupants, but also to passersby due to the shrapnel created by car parts. One infamous car bomb, actually transported in a rented truck, was that used in the bombing of the Murrah federal building in Oklahoma City, Oklahoma in 1995. In the United States., car bombs are investigated by the *Bureau of Alcohol, Tobacco, Firearms and Explosives (ATF)*. See *Oklahoma City bombing.*

carbon monoxide poisoning   the illness or death brought about by exposure to carbon monoxide, a colorless, odorless gas produced by the combustion of gasoline engines, gas stoves, camping lanterns, and other sources. Symptoms include dizziness, confusion, nausea, and vomiting.

career criminal   an offender who commits a variety of crimes over an extended period or who specializes in a particular type of crime, such as robbery. Career criminals have been the subject of legislation specifically designed to punish them more severely. Also referred to as *habitual offender*.

carjacking   the illegal seizure of a motor vehicle from its occupants, usually by means of force or threat of force. Carjacking resulted in the passage of legislation in many states designed to punish carjackers more severely. It has been suggested that carjacking came about as a result of increasingly sophisticated devices that prevent conventional *auto theft*.

carnal knowledge   to know another person in a sexual way.

carrying a concealed weapon   the practice of carrying a weapon, most often a handgun, in public in a concealed manner, either on one's person or in close proximity.

cartel   an organized criminal enterprise that often specializes in a particular type of crime, such as drug manufacturing and trafficking. See *Cali cartel, Medellin cartel, organized crime*.

cartographic school   an approach to the analysis of crime that makes use of geographical and social data. The cartographic school, which was prominent in the mid-19th century, is exemplified in the work of *Lambert Adolphe Jacques Quetelet* and *Andre-Michel Guerry*.

cartridge   a piece of ammunition consisting of a tubular container with propellant and the bullet or shot. Also known as round or shell.

case law   the law that emerges from interpretations by judges when deciding cases. This includes legal principles arising from decisions in appellate courts. As non-statutory law, case law is also known as *common law*.

caseload   the number of offenders that a probation, parole, diversion, or other officer supervises or refers for services. These professionals often cite high caseloads as a source of frustration because they inhibit the ability to effectively supervise offenders or render services.

caseload reduction   any program or strategy designed to reduce the number of criminal court cases awaiting disposition, especially those in danger of exceeding time limits. *Alternative dispute resolution (ADR)* is one means used for caseload reduction. Also referred to as docket reduction.

case study   an intensive examination of a specific set of circumstances in order to learn detailed information. Case studies yield more in-depth information than certain other research approaches, such as surveys.

C

cash bond   a *bond* that has to be posted in cash.

Castellammarese Wars   a series of bloody conflicts between Sicilian *Mafia* factions headed by Salvatore Maranzano and Joe Masseria in the early 1930s over control of organized criminal opportunities. See *organized crime.*

castoff   blood flung from a cutting or bludgeoning instrument as it is being used. Castoff helps investigators determine the method of death, the position of offenders and victims, the sequence of events, and whether others were present. See *blood spatter, bloodstain pattern analysis.*

castration   the removal or chemical neutralization of male testes employed in the treatment of chronic *sex offenders*. Physical castration involves the removal of the testes. In *chemical castration*, female hormones such as estrogen are used to compromise the primary sexual characteristics of offenders.

cat burglar   a thief who breaks into and enters residence or business by stealth, often at night by way of an upper story. See *breaking and entering, burglar.*

catharsis hypothesis   the assertion that violence as reported or portrayed in various media permits the audience to vent their aggressive tendencies and therefore experience an emotional release which then precludes their engagement in violent behaviors. Compare with *precipitation hypothesis.*

Caucasus Emirate   a Sunni organization committed to a global *jihad*. Compare with *Islamic State in Iraq and Syria (ISIS).*

causal fallacy   the notion that social policy intended to address crime is unconnected to the real causes of *crime.*

causation   see *etiology.*

cause of death   the injury or disease directly responsible for an individual's death. There can be a primary cause of death, such as a gunshot wound, and a related secondary cause of death, such as a massive brain injury. Compare with *manner of death.*

cautioning   in the United Kingdom, the warning of those who have committed minor offenses. Cautioning is not recommended for those who face serious, indictable offenses. Considerations in cautioning offenders include their age, their view of the *instant offense,* and their *prior record.*

celerity   the swiftness with which punishments for crimes are administered. Celerity is most closely associated with the *classical school of criminology*. Compare with *certainty, severity.* See *Bentham, Jeremy.*

cell   the self-contained room in a *lockup, jail, workhouse,* or *prison* where offenders are housed. Many cells include only a bunk, commode and wash facilities.

cellblock   an arrangement of cells within a *prison* or *jail*.

cell house   a building at a *prison* containing cells.

Center for Court Innovation   an organization based in New York and committed to undertaking innovative initiatives related to reducing crime and assisting crime victims.

Center for Criminological Studies   a center at the University of Oxford dedicated to pursuing research in criminology and to training graduate students.

Center for Evidence-Based Crime Policy   a unit within the Department of Criminology, Law, and Society at George Mason University that advances rigorous research into crime and justice policy.

Center for International Criminal Justice (CICJ)   a center at VU University in Amsterdam, The Netherlands that is dedicated to the study of *genocide*, *war crimes*, *crimes against humanity*, and *human rights violations*.

Center for Justice & Peacebuilding   a center at Eastern Mennonite University in Virginia that offers graduate training in conflict transformation. See *restorative justice*.

Center for Substance Abuse Prevention (CSAP)   a division of the U.S. Department of Health and Human Services responsible for promoting preventive approaches for substance abuse. CSAP makes grants to states, local, and private organizations.

Center for Substance Abuse Treatment (CSAT)   a division of the U.S. Department of Health and Human Services responsible for promoting treatment approaches for *substance abuse*. CSAT makes grants to states, local, and private organizations.

Center for the Study and Prevention of Violence (CSPV)   a center committed to conducting and disseminating research on the causes and prevention of violence. The CSPV maintains a literature database that is searchable on-line, and is located at the University of Colorado at Boulder. It publishes a list titled *Blueprints for Healthy Youth Development* that includes a number of prevention initiatives whose effectiveness has been demonstrated through rigorous evaluation.

Centers for Disease Control and Prevention (CDC)   a research and funding arm of the Public Health Service, U.S. Department of Health and Human Services. The CDC, as it is known, oversees an ambitious research agenda on a myriad of health problems, including homicide, suicide, and intentional injuries. It is also the sponsor of the periodic *Youth Risk Behavior Survey*.

Central Intelligence Agency (CIA)   the unit of the U.S. Government that gathers foreign intelligence information for purposes of national security. The CIA is headquartered in Langley, Virginia. See *espionage, spy*.

Central Park jogger case   the infamous assault and rape of Trisha Meili who had been jogging in Central Park in New York. Five young males were wrongfully convicted of the crimes. The actual perpetrator was eventually identified through *DNA testing*. The *convicted innocents* were later awarded $40 million. See *wrongful conviction*.

Centurion Ministries   a nonprofit organization based in Princeton, New Jersey dedicated to freeing and exonerating those who have been wrongfully convicted and imprisoned. See *Innocence Project, wrongful conviction*.

certainty   the aspect of a penal sanction that guarantees its imposition. See *classical school of criminology*. Compare with *celerity, severity*.

certified legal assistant   see *certified paralegal*.

certified paralegal   a *paralegal* who has completed a prescribed course of study and passed a qualifying examination.

chain gang   a form of punishment where groups of convicted prisoners are forced to perform manual labor while manacled together with iron chains. After having largely fallen into disuse in most jurisdictions, chain gangs saw a return in popularity with the movement toward *retribution* of the 1980s. See *Arpaio, Joe, manacles*.

chain of custody   the connection of *evidence* from one handler to another, from the point of collection to analysis to subsequent point in the *criminal justice system*. There are a number of ways a chain of custody can be broken. Defense attorneys use broken chains of custody to create *reasonable doubt* in the minds of jurors as to the integrity of the evidence, as well as the competence of law enforcement authorities.

chalk fairy   a police officer or other individual who feels compelled to draw a *chalk outline* around a homicide victim. Chalk fairies are considered a nuisance by homicide investigators.

chalk outline   a line drawn with chalk around a victim at a *crime scene* to show the location and position of a suspected homicide victim. Accepted law enforcement practice dictates that chalk lines should only be drawn when a body must be moved before the crime scene has been properly photographed, measured, sampled, and documented. See *chalk fairy*.

chamber   the compartment of a *firearm* housing the *cartridge* when the firearm is discharged.

Chambliss, William (1933–2014)   an American sociologist and criminologist who contributed to the *sociology of law* and *critical criminology*. Chambliss, who took his graduate training at Indiana University, spent the latter part of his career at George Washington University. His books include *Law, Order, and Power* and *Power, Politics, and Crime*.

change of venue   the practice of changing the location of a *trial*, primarily to avoid prejudicial *pretrial publicity* that might affect the outcome of the proceedings.

Chapin Hall   a research center at the University of Chicago whose mission is to improve the well-being of children and families.

charge   a statement of the offense that an individual is alleged to have committed.

charge bargaining   plea bargaining where the negotiations center around reducing the type and/or number of charges. It has been suggested that law enforcement officers and prosecutors may overcharge in recognition that the number or severity of charges will eventually be reduced. See *plea bargain, sentence bargaining.*

Chartered Society of Forensic Sciences (CSFS)   an international professional organization that provides forensic practitioners with opportunities for professional development. The CSFS is headquartered in Harrogate, North Yorkshire, England.

check fraud   the deliberate criminal act of writing checks on nonexistent accounts, closed accounts, accounts other than one's own, or on an account with insufficient funds to cover the check. See *bad check.* Compare with *check kiting, passing bad checks.*

check kiting   executing checks without sufficient funds. Compare with *passing bad checks.*

chemical castration   the use of Dep-Provera which contains female hormones to diminish the sex drive in male sex offenders. Discontinuation of the drug results in the restoration of the offender's normal sex drive. See *castration, sex offender.*

chemical imbalance   abnormal levels of certain chemicals in the body that can cause behavior changes and various mental disorders, such as depression. *Serotonin*, a neurotransmitter, is an example of such a chemical which can cause a type of imbalance.

Chessman, Caryl (1921–1960)   a convicted robber and rapist whose conviction drew attention from *capital punishment* abolitionists due to the questionable circumstances surrounding his *conviction* and his lengthy stay on *death row*. Chessman was executed in California's *gas chamber*.

C

Chicago Area Project   a large-scale social experiment conducted in disadvantaged neighborhoods in Chicago and designed to address *social disorganization* and its related symptoms. The Chicago Area Project was founded in the 1930s by *Clifford Shaw* and continues today. See *Chicago School of Criminology, ecological school of criminology.*

Chicago Gun Project   a project of Columbia Law School developed to use in-depth knowledge of social networks to reduce the incidence and transmission of gun violence in Chicago.

Chicago Jury Project   a series of in-depth studies of juries and jurors undertaken at the University of Chicago Law School in the mid-20th century. The Chicago Jury Project shed light on the jury process whose inner workings were previously unknown. See *grand jury, jury, petit jury.*

Chicago School of Criminology   a school of criminological thought and associated lines of research that had its beginnings in the Department of Sociology at the University of Chicago. The Chicago School was a product of several prominent intellectuals, including early sociologists Ernest Burgess and Robert Park and their students. The Chicago School, which focused on the *social disorganization* of urban areas, has influenced many criminologists.

Chicago Seven   seven individuals charged with inciting the disturbances at the 1968 Democratic National Convention in Chicago, Illinois. The Chicago Seven, defended by William Kunstler, included Rennie Davis, David Dellinger, John Froines, Tom Hayden, Abbie Hoffman, Jerry Rubin, and Lee Weiner. Black Panther Bobby Seale was one of the original defendants, but was tried separately due to his conduct in the courtroom. Several were convicted, but the convictions were overturned. See *civil disobedience, riot.*

Chicago typewriter   nickname for the *Thompson submachine gun.*

Chicago Violence Reduction Strategy   a program started in 2009 with funding from the MacArthur Foundation to bring law enforcement, social service agencies, and community members together to reduce gang-related shootings and homicides.

chief justice   in a federal or state supreme court, the *justice* who presides over the other justices. In some jurisdictions, the chief justice is elected to the position; in others, he or she is appointed. See *justice.*

Chikatilo, Andrei (1936–1994)   a Russian *serial killer* responsible for the deaths of 52 adults and children. The married father of two children, Chikatilo sexually assaulted, tortured, and mutilated his victims. He was found guilty of his crimes and executed by a single gunshot in 1994. See *serial murder, sexual homicide.*

child abandonment   the act of a parent deserting or illegally giving up his or her child.

child abduction   the abduction of a child.

child abuse   the intentional harm of a child by a parent or caretaker. Compare with *child neglect*.

child death review   the official reexamination of the circumstances surrounding the deaths of children in order to understand how and why they died, and to prevent future deaths. See *child homicide*.

child endangering   the intentional or unintentional behavior by a parent, guardian, or caregiver that places a child in danger.

child homicide   a homicide where the victim is a child. Child homicide differs substantially from adult homicides in that younger children tend to be killed at home by parents or caregivers, whereas older children are more likely to be killed away from home by friends or strangers. See *child death review*.

child molestation   illicit, unwarranted contact with a child, usually with sexual intent.

child molester   one who engages in *child molestation*. See *pedophilia*.

child neglect   the failure of a parent or caregiver to meet the basic needs of a child, such as food, water, clothing, and medical care. Research shows that child neglect can lead to a higher likelihood of *delinquency* and other problems with the child.

child pornography   *pornography* featuring children. See *Meese Commission*.

child protection order   a court order designed to protect a child from additional abuse from a parent or caregiver. Child protection orders typically require that the abusive person maintain a specified distance from the protected child. See *protection order*.

Children In Need of Supervision (CHINS)   youthful offenders whose needs are best served by supervision and services rather than by court processing or punishment. Compare with *Persons in Need of Supervision (PINS)*.

Children of Murdered Parents (CMP)   a nonprofit organization to provide supports to individuals whose parents were victims of homicide. Compare with *Parents of Murdered Children (POMC)*.

children who witness violence   a term used to describe youths who suffer physical, mental, or emotional trauma as a result of witnessing *homicide* and other acts of violence. Also, the term used by programs to identify and treat or conduct research on such behavior. In some cases, a crisis response team that is on call around the clock travels on site to assess such children and refer them for trauma services.

child saving   a movement in the United States that emphasized early intervention in the lives of children and resulted in the development of the modern juvenile justice system.

child sex trafficking   the *trafficking* of children for the purposes of sexual exploitation.

child sexual abuse   the sexual exploitation or assault of children by adults.

child witness   a juvenile who is a witness to a crime. Officials in some jurisdictions recommend efforts like videotaping the testimony of child witnesses in order to spare them the trauma of courtroom *testimony*. See *children who witness violence*.

Chinese Tongs   Chinese organizations that engage in a variety of criminal enterprises, including extortion. Compare with *Chinese Triads*.

Chinese Triads   Chinese secret organizations alleged to control *organized crime* in New York's Chinatown and other predominantly Chinese communities. Compare with *Chinese Tongs*.

CHINS   abbreviation for *Child in Need of Supervision (CHINS)*.

chokehold   a method of subduing a criminal suspect by tightly seizing the neck, generally from behind. This technique, occasionally used by police officers to restrain a suspect, has been the subject of controversy because of public concern over the excessive use of force by police, especially when the chokehold results in death of the *arrestee*. See *excessive force*.

chop shop   an illegal business that takes stolen motor vehicles apart in order to sell the constituent parts on the *black market*. Chop shops are highly profitable because auto parts sold separately are worth more than the intact vehicle.

chop wound   wound in human tissue made by the application of a heavy cutting instrument, such as a hatchet or machete. Compare with *incised wound*. See *sharp force trauma*.

chromatograph   a machine used to analyze the composition of chemical mixtures.

chromatography   the science of analyzing the composition of chemical mixtures.

chronic offender   a person arrested and convicted of serious crimes so often that they are deemed beyond *rehabilitation*. For such offenders, the concern of officials usually is public safety through *incapacitation* rather than rehabilitation, resulting in more severe sentences. Chronic offenders often are subject to longer, more severe prison sentences.

church arson   any attempted or completed burning or bombings of churches, particularly African American churches by white extremist groups or individuals. Instances of church arson have occurred more often in the Southern United States. See *Birmingham church bombing, hate crime.*

circle sentencing   a form of criminal sentencing derived from Native American practices where the offender permits members of the community to express their feelings about the crime and to make their requirements known. The name comes from the practice of participants arranging themselves in a circle. See *community justice, restorative justice, shaming penalties.*

circuit court   traditionally a court where judges preside in different locations for periods of time.

citizens academy   an education and prevention program designed to acquaint citizens with the duties and challenges of law enforcement.

citizen's arrest   the act of an ordinary citizen placing another person under arrest for a *felony*. While citizen arrests are possible and indeed desirable in the case of certain felonies, legal liability accrues to the individual assuming this power.

Citizenship and Immigration Services, United States (USCIS)   a division of the U. S. Department of Homeland Security that conducts inspections and investigations related to immigrants. Formerly called the Immigration and Naturalization Service (INS), their responsibilities include detaining and removing *illegal aliens*. The agency received much attention for its role in the aftermath of the *Attack on America.* See *Border Patrol.*

civil death   term used to express the loss of rights due to a felony *conviction.* See *felony disenfranchisement.*

civil disobedience   a form of non-violent protest where a person refuses to obey certain laws or regulations. See *campus unrest.*

civil liberties   personal guarantees and freedoms that the government cannot abridge, either by law or by judicial interpretation.

civil protection order   a court order designed to protect one person from another. Civil protection orders primarily are used in cases of *domestic violence*. Research on civil protection orders suggests that they not only keep most domestic violence victims safe from future abuse, but they also improve their sense of safety and well-being. See *internal affairs.*

C

civil remedy   other than an indictment in a criminal case, this is a remedy that the law gives to a victim against an offender. Examples of civil remedies include financial judgments.

civil rights   protections against the infringement of guaranteed freedoms. Some crimes can constitute violations of civil rights.

civilian review board   an appointed board whose role is to objectively examine allegations of *excessive force* and other forms of misconduct by police. Civilian review boards can be at the center of controversy because police often resent interference from outsiders.

clandestine drug lab   a makeshift, sometimes mobile laboratory that produces illegal drugs like *methamphetamine*. Clandestine drug labs can pose an *environmental hazard* due to the toxicity and volatility of the chemicals.

class characteristics   characteristics of evidence shared with other objects. Compare with *individual characteristics*.

classical school of criminology   a school of criminological thought of the late 18th and early 19th century emphasizing free will of those engaging in crime and proportionality between the offense and the punishment. Responding in large part to the arbitrariness and barbarity of the *inquisitorial justice*, the classical school favored the abolition of capital punishment and torture. The classical school of criminology, most closely associated with *Cesare Beccaria* and *Jeremy Bentham*, cast the individual as hedonistic, pursuing pleasure, and avoiding pain. Humans were considered rational beings, able to consider and choose among various courses of actions. This implied that officials needed only to devise punishments severe enough to deter those contemplating crime. Compare with *neo-classical school of criminology*.

classification instruments   paper and pencil instruments designed to assist correctional officials in determining inmate needs and problems and assigning them to appropriate institutions and programs. An example of a classification instrument is the *Level of Service Inventory-Revised*, which is a quantitative survey that enables correctional officials to assess both risks and needs of offenders.

clearance   the process of closing a criminal case by a suspect's arrest or death, or by some other means.

clearance rate   the rate of crimes cleared by arrest. Law enforcement agencies consider clearance rates an important statistic to demonstrate their efficiency and effectiveness.

clemency   an official order by a president or governor setting aside an *offender's* punishment for a crime. Clemency becomes controversial when those who have committed violent crimes are released, often as a final act of the official before leaving office. President Bill Clinton drew much criticism for granting clemency before the expiration of his second term.

clerk of courts a government official whose job is to file, maintain, and retrieve records pertaining to court cases.

Clery Act a federal act that requires colleges and universities to disclose information about campus crime. Institutions that fail to do so face penalties, such as losing various types of federal support.

Client-Specific Planning a process developed by the *National Center on Institutions and Alternatives* to tailor individualized sentencing and treatment plans for offenders convicted of non-capital crimes. Client-Specific Planning strives to identify a suitable sentencing alternative which incorporates the needs of both the offender and society. See *private presentence investigation reports.*

Clinard, Marshal B. (1911–2010) an American sociologist and criminologist of the mid- to late 20th century. Trained at the University of Chicago, Clinard spent most of his academic career at the University of Wisconsin. His contributions include pioneering work on *corporate crime*, a cross-cultural study of crime, and with Richard Quinney, the development of a criminal behavior *typology*. His books include *The Sociology of Deviant Behavior* and *Corporate Crime.*

clinical criminology the field that applies criminological findings to the assessment and treatment of offenders. In Israel, clinical criminologists are licensed by the ministry of health. See *Client Specific Planning, private presentence investigation reports.*

clinical prediction when clinical experience is used to formulate a prediction regarding an individual as "future dangerous," or "tendency to offend," by use of psychometric tests and other diagnostic tools of the clinician. Clinical prediction is criticized for its lack of reliability in determining the likelihood of *recidivism*. Compare with *statistical prediction.*

close security a level of security in prisons below *maximum security* and above *minimum security*. Prisoners placed in close security typically are escape risks or chronic rule violators. Compare with *maximum security, medium security, minimum security, trusty security.*

cloud the network of shared resources on the Internet. See *cyber crime.*

cloud computing the access and use of programs over the Internet. See *cyber crime.*

club drugs substances of abuse that enjoy widespread popularity among young people who frequent clubs, bars, and all-night dance parties known as raves or trances. Popular club drugs include *Ecstasy, Rohypnol, ketamine, methamphetamine* and *LSD*. See *substance abuse.*

Club Fed   nickname for the *Federal Law Enforcement Training Center (FLETC).*

Club, The   a steel bar with hooks fastened across the steering wheel of a motor vehicle to prevent *auto theft*. Car thieves have been known to cut through the steering wheel of a car to remove The Club, eliminating its effectiveness. See *crime prevention.*

clue   a piece of information that holds the potential of assisting in solving a crime.

cluster analysis   a statistical technique designed to identify groups of cases under study. Cluster analysis is used in criminology and criminal justice to statistically group different types, such as shoplifters, drunk drivers, batterers and child molesters.

clustered crime scene   a *crime scene* where most of the crime-related activity took place, including the confrontation, the attack, any sexual assault, and the homicide, if applicable. Compare with *primary crime scene, secondary crime scene.*

coagulate   to transform from a liquid to a semisolid. Blood coagulates into a blood clot.

cocaine   a white, crystalline alkaloid, $C_{17}H_{21}NO_4$, derived from cocoa leaves. Although it has medicinal benefits, it is a widely abused drug throughout the world. Colombia is a major supplier, the United States a major consumer. Drug cartels in Colombia primarily control the production and distribution of cocaine. See *crack cocaine.*

Cochran, Johnny (1937–2005)   a criminal *defense attorney* best known for his participation on the so-called *Dream Team* in the defense of *O. J. Simpson.*

Cochrane Collaboration   a network of health professionals committed to making synthesized research evidence available in order to better inform health decisions. Compare with *Campbell Collaboration.* See *evidence-based practice, systematic review.*

codefendant   a person charged with committing the same crime as a *defendant.*

Code of Hammurabi   the earliest written set of laws governing human behavior dating to the 18th century BC. Grounded in Babylonian religious beliefs, the Code of Hammurabi was considered humanitarian for its time. It is said that modern laws derive from the Code of Hammurabi.

code of the streets a term coined by sociologist Elijah Anderson to describe an oppositional culture in urban, socially disadvantaged neighborhoods characterized by informal rules governing behavior, including aggression and violence. See *social disorganization*.

CODIS abbreviation for *Combined DNA Index System*.

coerced confession a *confession* extracted from a criminal *suspect* involuntarily. Coerced confessions were more common before the *Miranda warning*. Such confessions are inadmissible in the United States, but may be admissible in other countries, some of which admit the confession as evidence, but punish the officer responsible for the coerced confession.

coerced treatment treatment of substance abuse, mental illness, or other problems that is mandated by judicial or administrative order. Also called coercive treatment.

coercion theory an *integrative theory* in criminology put forth by criminologist Mark Colvin that asserts that there are four potentially coercive environments that lead to criminal involvement: parents, peers, school, and neighborhood.

cognitive behavioral therapy a form of psychodynamic therapy that purports to alter dysfunctional thinking patterns in patients by presenting scenarios. Research has shown that cognitive behavioral therapy is beneficial in correcting thinking errors in offenders and can be an effective tool to better manage stressful life situations.

cohort study a group of persons who share something in common, such as year of birth. Criminologists study cohorts of people over time to assess their long-term involvement in delinquency and crime.

Cointelpro short for counter intelligence program. A *Federal Bureau of Investigation* initiative to infiltrate, disrupt and discredit liberal organizations. See *Black Panther Party*, *political crime*.

cold case a criminal case that has proven difficult to solve due to the shortage of leads or evidence. See *cold case unit*.

cold case squad same as *cold case unit*.

cold case unit a unit within a law enforcement agency that specializes in trying to solve cold cases. See *cold case*.

collar a slang term for an *arrest*. Also an individual who is arrested.

collateral consequences generally unanticipated consequences of a criminal justice policy or practice. An example is residency restrictions on sex offenders resulting in limited locations where they can legally live.

C

collective conscience   a term used by early sociologist *Emile Durkheim* to describe the beliefs held by a society's citizens. The collective conscience permeates society and is transmitted from generation to generation.

collective efficacy   the tendency of members of a neighborhood or community to look out for one another's interests, including serving as surrogate parents.

collective violence   violent behavior that results from many persons coming together, even where many of the participants would not have been violent if acting alone.

collector   a serial *sex offender* who collects souvenirs from victims, such as jewelry or articles of clothing.

collusion   a secret agreement between two or more people for the purpose of fraudulent or deceitful activity.

Colombian necktie   slang for a cut throat through which the deceased's tongue protrudes. Said to be used by Colombian drug dealers as a warning sign to others.

Columbine massacre   an infamous *mass murder* that took place in April 1999 at Columbine High School in the Denver, Colorado suburb of Littleton where 15 people died. Columbine students *Eric Harris* and *Dylan Klebold* entered the school armed with semiautomatic weapons, shotguns, and homemade bombs. After fatally shooting 12 of their fellow students and a teacher, as well as wounding others, Harris and Klebold took their own lives. See *lockdown, mass murder, Newtown massacre, school violence, semiautomatic weapons.*

Combined DNA Index System (CODIS)   a system designed to facilitate the exchange and comparison of DNA evidence from multiple jurisdictions in the United States and abroad.

commercial burglary   the *burglary* of a commercial business. Compare with *residential burglary.*

Commission on Accreditation of Law Enforcement Agencies (CALEA)   a credentialing organization whose purpose is to improve law enforcement agencies by promulgating standards and recognizing excellence through accreditation. CALEA is headquartered in Gainesville, Virginia.

commitment   one of four elements of Travis Hirschi's control theory of delinquency. See *attachment, belief, involvement.*

commodity fraud   *fraud* involving the trading of goods or merchandise.

common law   the criminal law as interpreted through cases and precedent throughout the ages.

common pleas court   in some states in the United States, courts having original jurisdiction in criminal matters, unless the cases derive from an inferior court, such as a county or municipal court.

Communities That Care   a comprehensive process for community change designed to reduce delinquency and promote healthy youth development. Communities That Care was developed by the *Social Development Research Group* affiliated with the University of Washington.

community aid panel   a group consisting of a *police officer*, a *solicitor* and responsible community members who assist an *offender* in addressing the problems which led to their illegal behavior. Used in New South Wales, community aid panels begin their efforts at *rehabilitation* prior to *sentencing*.

community-based correctional facility (CBCF)   a residential center specifically designed to permit convicted offenders to remain in the community. CBCF residents may receive appropriate drug or other treatment, may work regular jobs, and may participate in other activities that are often not part of institutional corrections.

community-based corrections   see *community corrections*.

community-based sanctions   sanctions that permit the offender to remain in the community. Community-based sanctions can include *probation*, fines, residence in a *halfway house*, *electronic monitoring*, or other alternatives to confinement.

community beat officer   a law enforcement officer that proactively works with citizens and businesses in a specific area to maintain public safety.

community control   an umbrella term for a sentencing option when the offender is not in prison or jail, but still under control. These controls can include residential arrangements, such as a *halfway house* or supervision on *probation*.

community corrections   the spectrum of sentencing alternatives that permit the convicted offender to remain in the community as opposed to serving time in a remote correctional facility. Community corrections include, but are not limited to, community-based correctional facilities, halfway houses, day reporting centers, *probation*, and *parole*.

community court   a *specialty court* where lay members of the community participate in developing a just disposition for the *offender*. See *circle sentencing, restorative justice*.

community justice   a movement that began in the late 20th century to involve the community in the disposition of criminal cases, based on the premise that the community both gives rise to criminality and suffers the consequences and therefore should participate in its resolution.

community oriented policing   same as *community policing.*

community policing   a philosophy of policing that emphasizes identifying and solving a wide range of community problems that are thought to lead to crime and social disorder. In community policing, often simply shortened to community policing, the beat officer and community resident form a bond that permits the regular exchange of information to promote safety and improve the overall quality of life in the neighborhood. Community policing is closely associated with the late *Robert Trojanowicz* of Michigan State University. Compare with *problem-oriented policing.*

community prosecution   the practice of strengthening the relationship between district attorneys and the communities they serve. Starting in the early 1990s, community prosecution programs took the lead in *community policing* by connecting criminal justice services with neighborhoods. See *prosecuting attorney.*

community service   a sentencing alternative where the offender performs work for the public good. Community service includes such tasks as picking up trash, painting public buildings, and removing graffiti. Those sentenced to community service generally have to abide by rules, such as showing up for work on time, not taking substances of abuse, and performing the required service safely.

community service order   an order issued by a court specifying that an offender participate in *community service.* Community service orders are often part of conditions of *probation.*

community service restitution   volunteer work for the purpose of giving back to the community as a result of a criminal conviction. See *community service, restitution.*

community standards   standards of decency dependent on what local citizens consider acceptable. See *adult bookstore, pornography.*

commutation   the process of changing a criminal sentence to make it less severe. A commutation can only be granted by executive authority. Compare with *pardon.*

comorbidity   the state of having two or more diseases or conditions at the same time. An example regarding crime is the offender who has both a *substance abuse* and a mental health problem.

compact   an agreement between jurisdictions to accept and supervise convicted offenders. For example, an interstate compact can permit probationers or parolees to relocate to other states with the understanding that supervision will continue after they move.

comparative criminology the study of crime and criminal justice processes across other countries and cultures.

comparative justice the study of criminal justice systems across other countries and cultures.

comparative policing the study of police and policing across other countries and cultures.

comparison microscope a microscope that enables the analyst to compare two samples simultaneously. Uses of a comparison microscope include comparing two bullets.

comparison prints fingerprints taken from possible suspects or others to be compared with *latent prints* found at a *crime scene*. Compare with *elimination prints*.

compensation order in Britain, an order to make restitution instead of or in addition to a *fine*.

competency hearing a court hearing held to establish if a *defendant* is competent to stand *trial*.

competent capable of understanding the nature of the crime or crimes for which one has been accused.

complainant one who swears out a criminal *complaint* against another.

complaint a formal charge alleging that an individual has committed a particular offense on a particular date.

completed crime a crime which has been carried out. The *severity* of punishment for a completed crime generally is greater than for an *attempted crime*.

complicity the association or participation in serving as an accomplice in a crime.

composite a picture of a criminal suspect, produced either manually or by computer. For several decades, law enforcement agencies employed artists to render sketches of suspects based on information provided by witnesses. More recently, sophisticated computer programs permit police to combine a wide variety of facial features to achieve a similar result. Another advantage of the latter is the ability of non-artistic lay persons to develop composites more rapidly.

composite drawing an artist's rendering of an offender, missing person, or other individual of interest in a criminal case. See *composite, forensic artist, Identikit*.

compounding a felony agreeing to refrain from prosecution in return for financial or other material consideration. Compounding a *felony* is illegal.

C

Compstat   a process that originated in New York that uses a wide variety of data to map crime in order to improve the quality of life for citizens. Compstat evolved into a data-driven management computer program.

computational forensics   the application of computer-based modeling to improve *forensic science.*

computer crime   a criminal offense facilitated or made possible by the use of a computer. Examples of computer crimes include the electronic transfer of funds. See *computer virus, hacker, phishing.*

computer criminal   a person who uses a computer to commit a crime. See *computer virus, hacker.*

computer forensics   a forensic specialty that focuses on the retrieval and analysis of *evidence* from computers and from the *cloud.*

Computer Fraud Abuse Act (CFAA)   a federal anti-hacking law that prohibits unauthorized access to a computer. See *hacking.*

computer virus   a computer program designed specifically to annoy the user or cause damage to the computer or its programs or data. Despite the lack of criminal intent in some of these cases, they can be responsible for millions of dollars in damage or other financial losses. See *computer crime, hacker.*

computerized criminal histories (CCH)   electronically stored and retrievable records of offenders' arrests and case dispositions. CCH is important for a variety of reasons, including the ability to check the prior records of those applying to purchase firearms.

con   short for *convict.* Also, short for *confidence game.*

concealed carry permit   a permit issued by state or local government allowing an individual to carry a concealed firearm, most often a handgun.

Concealed Information Test   a method of uncovering the deception of a criminal suspect by focusing on details that only the actual offender could know. The Concealed Information Test avoids problems associated with stress experienced by innocent persons.

concealment of birth   hiding the birth of a child. See *neonaticide.*

concealment of death   hiding the death of a person. Deaths are sometimes concealed in order to commit fraud, such as when a family member continues to cash the deceased relative's Social Security or retirement checks.

concentrated disadvantage   the spatial density of socioeconomic disadvantage. Areas of concentrated disadvantage have fewer opportunities and greater challenges, including delinquency and crime. See *Chicago School of Criminology.*

concentration camp a camp where prisoners of war or political prisoners are held. See *Holocaust*.

concentric zone theory a sociological approach to the analysis of crime, delinquency, and social disorder that lays out cities as a series of concentric rings or zones, with the central city represented by the inner ring. The concentric rings move outward to the suburbs. Each circle represents a different phase of industrial and socioeconomic development. The concentric zone approach is most closely identified with early University of Chicago sociologists Ernest W. Burgess and Robert Park. See *Chicago School of Criminology*.

concertina wire coils of wire, often equipped with sharp barbs, affixed to the tops of prison or jail walls in order to prevent escape or unauthorized entry. Compare with *razor ribbon*.

conciliation indirect mediation between a victim and the offender. See *victim-offender mediation*.

concurrent sentences sentences imposed by a criminal court that are served at the same time. Compare with *consecutive sentences*.

conditional discharge a disposition which permits the dismissal of a criminal case if certain conditions are met.

conditional release the release of a *defendant* by the court under certain conditions, such as participation in a treatment program. See *unconditional release*.

conditions of confinement the living conditions in a detention or correctional facility.

conditions of release special requirements of *probation* or *parole*.

conduct disorder a disorder characterized by a disregard of social norms. Conduct disorder is a common diagnosis for a youth who has engaged in violent or other antisocial behavior. Compare with *antisocial personality disorder*.

conductive electrical weapon a weapon whose mechanism employs high voltage to temporarily immobilize a person. Tasers and stun guns are examples of conductive electrical weapons. See *stun gun, Taser*.

conduct norms unwritten rules that guide how people should act. Conduct norms that are developed over time through the interaction of group members can vary from group to group. Some conduct norms eventually find expression in the passage of laws. Compare with *folkways, mores*.

confession an acknowledgment of involvement in a crime, made either verbally or in writing. See *coerced confession*.

C

confidence game   a crime where the offender elicits the trust of the victim, often playing on sympathy or greed, to relieve the latter of money or other goods.

confidence man   one who operates a *confidence game*.

confidential informant (CI)   an individual who provides confidential information to law enforcement officials about crimes under investigation, sometimes in return for financial or other consideration.

confidentiality   the principle and practice of not revealing the information shared by one party with another.

confinement   the condition of being locked up in either a short-term or long-term facility for offenders.

confirmatory test   in forensic science, a test of a sample that determines what it is not as well as what it may be. Compare with *presumptive test*.

confiscated assets fund   a public account whose deposits derive from confiscated criminal assets.

conflict model   a theoretical position in criminology which holds that opposing political, social, or other forces in society are responsible for a variety of social ills, including crime and delinquency. Inherent in the conflict model is that there is little normative consensus. Also referred to as *conflict theory*.

conflict of interest   a situation where a party to a legal matter or business transaction cannot participate due to a preexisting personal or professional relationship with another party.

Conflict Tactics Scale   created by sociologist Murray Straus, a widely-used scale that assesses how individuals and their partners deal with conflict.

conflict theory   see *conflict model*.

conjugal visit   an unsupervised visit of a spouse or significant other to a partner who is confined in prison or jail for the purpose of facilitating sexual relations. Conjugal visits for prisoners have been common in Scandinavian countries for many years.

con man   same as *confidence man*.

consecutive sentences   sentences served one after the other. Criminal defendants typically receive consecutive sentences in the case of particularly serious crimes. Compare with *concurrent sentences*.

consensus model   a model in criminology which posits that most people in society subscribe to the same set of norms and values. Compare with *conflict model*.

C

consent decree   a decree issued by a judge that expresses the agreement between two or more parties to a dispute.

consigliere   in an *organized crime* family, one who serves as adviser to a *boss*. The role of *consigliere* was dramatically portrayed by actor Robert Duvall as Tom Hagen in the film *The Godfather*.

conspiracy   a secret, organized effort by two or more individuals to commit a crime. Persons may be charged with conspiracy even though the planned act is never completed.

conspirator   one who willingly participates in a *conspiracy*.

constable   a *peace officer* with limited authority and jurisdiction. In Britain, a police officer.

constitutional theory   any theoretical perspective in criminology that focuses on the physical, biological, or mental constitution of the person. See *Cesare Lombroso*.

consumer fraud   *fraud* perpetrated against those who purchase merchandise or services. An example of consumer fraud is a person performing shoddy house repair after receiving payment for proper repairs.

contact wound   a wound that occurs when a firearm is discharged while in direct contact with the body. The infusion of gases causes a violent bursting of the skin and a wound with poorly defined edges. See *entrance wound*, *starring*.

containment theory   a form of *control theory* advanced by American criminologist *Walter C. Reckless* which posits that individuals are insulated from the risk factors of crime and delinquency by inner containment and outer containment. Inner containment consists of an individual's self-concept and subscription to conventional values and goals. Outer containment consists of the formal and informal strictures placed on the individual by laws and norms. See *good boys, bad boys*.

contamination of evidence   adulteration of *evidence* by the introduction of a foreign substance, either deliberately or accidentally. Contamination of evidence can result in the inability to successfully pursue charges against an accused.

contempt of court   a disregard for the respect and decorum of a court of law. Also, the charge for such disregard. Contempt of court can result in the confinement of the person found in contempt, as in the cases of reporters who invoke the *First Amendment* to protect their sources.

content analysis   a set of qualitative data analysis techniques that permit the analysis of text for terms, concepts, themes, or theoretical constructs.

Content analysis has been used in criminological research to examine newspaper coverage of certain types of crime.

continuance   a postponement of a hearing, trial, or other court appearance. Continuances may be sought to delay proceedings or simply to annoy the other party.

continuum of care   theoretically, the seamless delivery of services to delinquents or children at risk. A continuum of care improves the coordination and effectiveness of such services.

contraband   prohibited goods, such as those acquired by jail and prison inmates.

contract   an offer to pay for the killing of a person. See *hit, hit man.*

contrecoup contusion   a contusion of the brain found opposite the point of impact. Contrecoup contusions result from the brain moving back and forth within the skull.

control group   in experimental research, the group identical to the experimental group in all respects except for the fact that this group receives no treatment.

controlled substance   any narcotic or other drug whose manufacture, sale, or distribution is regulated by federal law.

control theory   a set of theoretical perspectives in criminology which assert that certain social forces serve to prevent delinquency and crime by controlling individuals who otherwise would pursue self-interest. Control can be either internal or external. Examples of control theories include *containment theory* and a *General Theory of Crime.*

contusion   a bruise caused by a blow or impact.

convergence hypothesis   the notion that as males and females move toward equality, the differences between them in the amount of criminal activity will lessen. So, as women attain greater sex role equality as measured by such factors as participation in the workforce and the amount of education they possess, it is expected that they will engage in more crime, especially *property crime.*

conversion   the act of illegally appropriating the property of another for one's own use.

conversion fraud   taking a person's money by deception and using it for one's own purposes. An example of conversion fraud is a child who uses access to their elderly parent's bank account to fund their own activities.

convict   to prove someone guilty of a crime. Also, a person who has been found guilty of a crime, particularly one who is serving or has served time in prison.

convict criminology a branch of criminology that questions traditional criminology's views of crime and offenders that are uninformed by the realities of imprisonment. See *felony disenfranchisement*.

convicted innocent an individual who was wrongfully convicted of a crime. See *wrongful conviction*.

conviction a finding of a defendant's guilt by a judge, jury, or by a guilty plea.

conviction record the official documentation of an offender's record of convictions. See *conviction*. Compare with *arrest record*.

cooling-off period the spans of time between the offenses of a *serial killer*. Also, the period between when someone applies to purchase a firearm and when they actually take possession of it; this period is intended to permit any anger on the part of the purchaser to subside. See *serial murder*.

cop-killer bullets term given to Teflon-coated bullets capable of penetrating the typical *bulletproof vest*. Cop-killer bullets are prohibited by law.

copycat a person who copies the *method of operation* of a publicized crime.

copyright infringement unlawful reproduction or use of copyrighted material.

Corcoran Eight eight California correctional officers who were federally charged with depriving inmates of their *civil rights* by encouraging the inmates to fight.

coroner a public official charged with the responsibility of investigating all deaths other than from natural causes. In most jurisdictions, coroners are elected. Their authority to investigate the circumstances surrounding crimes supersedes that of all other local officials. Coroners need not have formal medical training. Compare with *medical examiner*.

corporal punishment punishment that generally consists of a form of hitting or beating. Once common in families and schools, corporal punishment has waned largely in response to lawsuits by parents and with the belief that violence begets later violence.

corporate crime crime perpetrated by a corporation's officials for its benefit. Examples of corporate crime include the production and sale of dangerously defective Pintos by the Ford Motor Corporation and the dumping of harmful chemicals in the Love Canal by Dow Chemical. Compare with *white-collar crime*.

corporate psychopath a *psychopath* who is attracted to corporate settings by the promise of power, financial gain, and control.

corpse a dead body, especially of a human being.

corpus delicti a Latin term for a body that can be seen and possessed. At one time, a corpus delicti was necessary in order to pursue homicide charges against a *suspect*.

correctional facility a structure where convicted offenders are housed as a result of a criminal conviction.

correctional officer see *corrections officer*.

corrections the field concerned with the imprisonment, control, and rehabilitation of convicted offenders. Corrections includes the administration and study of prisons, community-based sanctions, *parole*, and *probation*, as well as less intrusive alternatives.

Corrections Corporation of America a private company headquartered in Nashville, Tennessee that owns and operates correctional facilities in the United States.

corrections officer one who guards prisoners in a correctional facility. The role of corrections officer is fraught with controversy and ambivalence; the officers face the dangers of guarding violent offenders, but frequently are found guilty of engaging in brutality against inmates or of smuggling contraband into prisons. In most jurisdictions, corrections officers have strong union representation. Compare with *warden*.

correlation a statistical association between two variables, as in the relationship between socioeconomic status and crime.

corruption the widespread use of *bribery* or *fraud*.

corruption of a minor a sex offense involving intercourse or other sexual conduct by an adult against someone under the age of 17. Compare with *statutory rape*.

Cosa Nostra see *La Cosa Nostra*.

cost overrun the practice of contractors where they knowingly exceed estimated costs for merchandise or services in order to increase their profits.

Council of State Governments (CSG) a national nonprofit organization headquartered in Lexington, Kentucky that serves all three branches of state government through the exchange of ideas and information in order to improve public policy. The CSG maintains a Justice Center that offers advice and strategies to promote public safety and strengthen communities.

counselor a term for a lawyer who gives advice about the law. Compare with *barrister*, *solicitor*.

counterfeit illegally produced. Counterfeit money, for example, is currency that is illegally manufactured.

counterfeiting the process of producing *counterfeit* currency or other goods.

counter-narcotics law enforcement efforts devoted to fighting narcotics trafficking.

counterterrorism law enforcement efforts devoted to fighting terrorist activities.

country club prison a prison that offers amenities not typical of correctional facilities, such as comfortable rooms, swimming pools and tennis courts.

county court a *court* with jurisdiction over misdemeanors and traffic offenses in a community not having a municipal or *mayor's court*. County courts generally have jurisdiction over the entire county where they are located.

court a government institution whose responsibility is to adjudicate disputes and criminal charges.

court administrator the official whose responsibility is to oversee the operation of a *court*, including budget management, human resources, and docket management.

court appointed attorney an attorney assigned by a judge to represent an indigent defendant. Court appointed attorneys are frequently used as a supplement to, or instead of, a *public defender*. See *appointed counsel*.

court appointed special advocate (CASA) individuals appointed by a court to protect abused or neglected children. See *child abuse*.

court backlog the glut of cases awaiting disposition in a criminal court, some of which may be nearing the legal limits for trial. See *caseload reduction, court delay*.

court clerk a person responsible for maintaining the records of a criminal court. Court clerks often are elected to office. Also referred as *clerk of courts*.

court costs a fee assessed on convicted offenders by a *judge* or *magistrate* to help recover the costs associated with court processing.

court delay the delay in processing criminal cases in court. In order to address court delay, some jurisdictions have adopted time standards for the processing of criminal cases. See *court backlog, speedy trial*.

court liaison service a program in Australia that provides psychiatric assessment and intervention services to those before the Magistrate's Court.

court of appeals a court which hears appeals of lower courts.

court of public opinion an imaginary tribunal of lay persons whose judgment about a criminal case or other publicized matter is said to have a bearing on the case.

court reform   a term used to describe any of a wide variety of movements to change the way that courts are organized or the way that they operate. Court reform can include such issues as *court unification*.

court reporter   a specially trained stenographer who uses a transcribing machine to record the verbal proceedings in a court of law. In some jurisdictions, court reporters are employed by the court. In others, they are self-employed. Court reporters also transcribe the output from the transcribing machine, thereby making the transcripts available for purposes, such as appeals.

court stenographer   a stenographer employed or contracted by a court in order to produce a verbatim record of the proceedings. See *court reporter*.

Court TV   a television network debuting in the early 1990s that broadcasts a variety of crime-related programming. Court TV aired the widely viewed flight of *O. J. Simpson* and his subsequent trial. After covering well-publicized criminal cases, Court TV has evolved into truTV.

court unification   the placement of lower courts under the control of a centralized administrative umbrella. This can include, but not necessarily be limited to, centralization of authority, rulemaking powers, budgeting, and funding.

court watcher   a lay person who attends court hearings for the purpose of monitoring how justice is dispensed.

cover-up   the concealment of wrongdoing. Examples of cover-ups include efforts by police to mask the real circumstances of the questionable shooting of a suspect, including the use of a *throwaway* weapon.

CPTED   abbreviation for *Crime Prevention Through Environmental Design*.

crack   same as *crack cocaine*.

crack baby   an infant born addicted to crack cocaine as a consequence of its mother's addiction. Crack babies experience a myriad of problems, such as low birth weight and withdrawal symptoms if the drug in question is not administered.

crack cocaine   a crystalline form of *cocaine* usually smoked in a pipe. Crack cocaine was responsible for an unprecedented rise in crime in the 1980s and 1990s. Also referred to as *crack*.

crackdown   a policy of aggressive enforcement or prosecution of certain criminal offenses.

crack house   a house or other dwelling, often vacant, used by crack addicts to get high. Crack houses are often the target of law enforcement efforts to clean up socially disorganized neighborhoods.

crack whore a person who engages in prostitution to support a *crack cocaine* habit.

Craigslist killers nickname given to offenders who use Craigslist, a classified advertisements website, to lure the victims they meet and then murder.

craniofacial identification the science and technology associated with the detailed identification of the face and skull.

craniofacial reconstruction the science and art of reconstructing the head and facial features of a deceased person.

crank slang for *methamphetamine.*

creative accounting accounting practices that meet accepted standards, but violate ethical or moral principles.

credit card crime *theft* or *fraud* committed using or involving a credit card.

cremains human remains after cremation.

Cressey, Donald R. (1919–1987) an American criminologist best known for studying with and co-authoring with *Edwin H. Sutherland* their textbook *Principles of Criminology,* which was published in multiple editions over the course of 40 years. Among Cressey's contributions were a pioneering study of embezzlement and a monograph on organized crime. Cressey spent much of his career at the University of California at Santa Barbara.

crime behavior prohibited by law and punishable by a term of confinement, the imposition of fines, or other legal sanctions.

crime analysis the processing of data in order to shed light on the incidence, prevalence, or seriousness of crime and related behaviors.

crime boss the head of a criminal organization, such as a *crime family* or *drug cartel.*

Crime Classification Manual a written guide on violent crime developed by former *Federal Bureau of Investigation* profilers John Douglas and Robert Ressler and nursing professor Ann W. Burgess. The Crime Classification Manual details how to investigate and classify serial and other serious crimes.

crime clock heuristic device used by the *Federal Bureau of Investigation* in its *Uniform Crime Reports* to convey approximately how frequently each day each type of serious crime occurs. The use of the crime clock is deceptive because it implies that crime is evenly distributed throughout the 24 hours of a day.

crime commission a body created by legislative or administrative action whose role is to study crime and formulate recommendations for its prevention and control.

crime control any effort to curb the incidence and prevalence of criminal behavior.

crime czar an official who heads up federal, state, or local efforts to combat crime. Compare with *drug czar*.

crime displacement the movement of crime from one area to another due to law enforcement or other intervention efforts. Crime displacement is a consideration in determining the effects and effectiveness of *crime prevention* programs.

crime family a closely knit group of *organized crime* affiliates, some of whom are related, that engages in criminal enterprise. See *Mafia*.

crime fiction fiction whose plot revolves around the perpetration and solution of a crime, most often a homicide. Crime fiction encompasses several genres of fiction, including *mystery, suspense novel,* and *thriller*.

crime fighter a public servant or lay person who spends their time and resources preventing and controlling crime.

crime forecasting the practice of using statistical techniques to predict future trends. However, accurate crime forecasting can be difficult since future trends do not always mirror past trends. For example, the incorrect prediction of a wave of super predators that would victimize the United States in the early 21st century. No such crime wave subsequently occurred. See *ARIMA*.

crime index according to the *Uniform Crime Reports*, the total of the most serious crimes, including murder and non-negligent homicide, rape, robbery, aggravated assault, larceny, motor vehicle theft, and arson. The crime index is used by the *Federal Bureau of Investigation* as a general barometer of how much crime there is in the United States. See *modified index, Part I offense, Part II offense*.

crime laboratory a scientific facility designed to analyze *evidence* in criminal cases.

crime line a telephone line citizens can use to anonymously report crimes in return for financial rewards.

crime mapping the practice of plotting crimes on a map, most often using computer software. Crime mapping permits the visual inspection of crime patterns, which can inform law enforcement strategies.

crime novel a fictional book where the solving of a crime is central to the plot. See *mystery, suspense novel, thriller, whodunit*.

crime of commission a crime where the offender intends to perpetrate the act.

crime of omission a crime that results from the *offender's* failure to perpetrate the act.

crime of opportunity a crime committed by an offender that was ancillary to the intended offense. An example is a *rape* committed during the commission of a *burglary*.

crime of passion a violent crime spawned by extreme emotion, such as jealously or rage. See *expressive crime*.

crime pattern theory the theory in criminology that crimes occur when offenders and victims converge in time and space.

crime prevention the specialized field of study and practice concerned with keeping crime from occurring in the first place. Crime prevention encompasses efforts to reduce opportunities for victimization as well as those intended to avert youths from delinquency. See *National Crime Prevention Council (NCPC)*.

Crime Prevention Through Environmental Design (CPTED) the title of a book by criminologist *C. Ray Jeffrey*, as well as an area of specialization within the field of *crime prevention*. The advocates of CPTED maintain that by eliminating the opportunities for crime inherent in the physical environment, many crimes can be prevented. CPTED was originally inspired by the work of noted architect R. Buckminster Fuller. See *environmental criminology*.

crime rate the number of crimes per unit of population; the most common unit used is 100,000. The crime rate per 100,000 is calculated by dividing the actual population of the jurisdiction in question by 100,000. That quotient is then divided into the number of actual crimes to yield the crime rate. See *Uniform Crime Reports*.

crime reporter a journalist who specializes in covering stories related to crime and justice. See *Criminal Justice Journalists*.

crimes against humanity crimes so egregious they offend the sensibilities of most civilized societies. Examples of crimes against humanity are *genocide* and *war crimes*.

crime scene the physical area where a crime has taken place. In cases of violent and other serious crimes, great care is usually taken to cordon off the crime scene in order to preserve evidence which could lead to the identification, arrest, and conviction of the perpetrator.

crime scene cards cards placed at the scene of a crime to announce that the area is restricted to authorized personnel.

crime scene contamination activities, most often unintentional, which destroy or alter physical evidence or otherwise compromise the value of the evidence for prosecution purposes.

crime scene investigation (CSI) the meeting point of science, logic, and law involving purposeful documentation of the conditions at the scene and the collection of any physical evidence that could possibly illuminate what happened and point to who did it.

crime scene investigator one who engages in *crime scene investigation.*

crime scene reconstruction assembling the puzzle of a crime using available physical, spatial, and other information about the crime.

crime scene tape yellow plastic tape used to cordon off a crime scene. Crime scene tape is usually printed with a warning, such as "Police Line—Do Not Cross" or similar verbiage.

crime spree a series of criminal offenses committed by one or more offenders within a limited period of time. See *Starkweather, Charles, Starkweather syndrome.*

Crime Stat a computer program designed to analyze crime incidents in relation to their locations.

Crime Stoppers an initiative designed to permit the public to anonymously provide information about unsolved criminal cases, sometimes in exchange for monetary rewards.

Crime Survey for England and Wales periodic surveys of, and interviews with, residents in England and Wales regarding their experiences with crime.

crime trend the extent to which crime is going up, going down, or staying the same.

crime wave a period or trend of a higher amount of crime. Crime waves can be due to a variety of reasons, including the popularity of a new drug like *crack*, or the demographic effects of a *baby boom*.

criminal one who has committed a *crime.*

criminal alien an *illegal alien* who has committed, or is suspected of having committed, a crime.

criminal anthropology the association of body types and other physical characteristics with the tendency toward engaging in crime. See *Cesare Lombroso, positive school of criminology.*

criminal assets the ill-gotten money and other property acquired illegally.

criminal behavior system a *typology* of crime associated with criminologists Marshall Clinard and Richard Quinney. Their system included the following types of crimes: violent personal; occasional property;

occupational; corporate; political; public order; conventional; organized; and professional.

criminal career the longitudinal sequence of an individual's involvement in criminal behavior. The examination of criminal careers can include such issues as the age at which they begin to offend as well as the age at which they desist. Also of interest is whether criminals specialize in certain types of crimes or if they participate in a variety of offenses.

Criminal Cases Review Commission the commission that investigates possible wrongful convictions in England, Wales, and Northern Ireland.

criminal intent see *intent*.

criminal investigative analysis the application of in-depth knowledge about a suspect's personality and behavior in order to identify and apprehend predatory and other violent offenders. Central to criminal investigative analysis is the development of profiles based on comprehensive analysis of the crimes and the *offender's* behavior. See *Behavioral Analysis Unit (BAU), profiler.*

criminal investigative analyst see *profiler*.

criminalist a scientifically trained person who specializes in the detection or solution of crime through the analysis of physical evidence, such as DNA, fingerprints, tool marks, body fluids and other forms of evidence. See *DNA testing, forensic science.*

criminalistics a field concerned with the scientific investigation and detection of crime.

criminality the quality of being criminal or having such characteristics.

criminal justice funnel a heuristic device to show the large number of suspected or reported crimes and their attrition as they pass through the various stages of the *criminal justice system*. The large neck of the funnel represents all offenses.

Criminal Justice Journalists an organization of journalists that specialize in investigating and writing about crime-related topics. Criminal Justice Journalists is headquartered at the University of Pennsylvania. See *crime reporter*.

criminal justice system the entire governmental apparatus that formally processes crime, including, but not limited to law enforcement, prosecution, defense, the courts, and corrections.

criminal law the body of statutes that covers acts defined as criminal.

criminal liability responsibility for behavior that is against the law and results in harm.

criminaloid   according to the 19th century Italian criminologist *Raffaele Garofalo*, an offender who is motivated by emotion, in combination with other factors, commits criminal behavior. See *positive school of criminology*.

criminal opportunity   access to victims. Compare with *means, motive*.

criminal profiler   see *profiler*.

criminal sanction   a penalty for a crime. Typical criminal sanctions include imprisonment, fines, probation, and community service.

criminal trespass   a criminal offense, most often a *misdemeanor*, which involves the uninvited presence of an intruder. Compare with *breaking and entering, burglary*.

criminogenic factors   factors responsible for spawning crime, such as association with delinquent peers and faulty thinking patterns. Effective correctional programming targets criminogenic factors.

criminological theory   any systematic attempt to explain the causes or *etiology* of crime. See *theoretical criminology*.

Criminological and Victimological Society of Southern Africa   a nonprofit organization promoting the discipline of criminology in southern Africa.

criminologist   a person whose professional identity revolves around the study of crime, criminals, and the criminal justice system. Some people mistakenly call criminologists those who engage in the scientific detection of crime. Compare with *criminalist*.

criminology   the scientific study of crime, criminal, and the criminal justice system. Criminology encompasses not only the *etiology* of crime, but also the processing of offenders by law enforcement, prosecution, courts, and corrections.

Crips   a large Los Angeles, California based *gang* with a reputation for violence. See *Bloods, Latin Kings*.

crisis intervention   any of several services intended to assist persons in extreme emotional distress, including those contemplating suicide.

crisis intervention team   a team of specially trained law enforcement officers and mental health professionals that respond to situations involving mentally ill persons.

crisis negotiation   the process of carefully communicating with parties during a crisis with the end of peaceful resolution.

crisis negotiation unit (CNU)   a law enforcement unit that specializes in *crisis negotiation*.

critical criminology a criminological perspective that questions the right of those in power to define and control behavior as criminal. Critical criminology was in part an outgrowth of the *labeling perspective.* Compare with *Marxist criminology.*

critical incident analysis the in-depth analysis of highly stressful traumatic events, such as acts of terrorism, natural disasters, and school shootings.

critical incident response group a law enforcement unit that specializes in responding to critical incidents.

cross examination after a witness has testified in court, the interrogation of that witness by the opposing side.

crossover youths youths who are involved in both the juvenile justice and child welfare systems. As such, crossover youths have more problems that call for a broader spectrum of services.

cross-sectional study a research study that takes a one-time snapshot view of the phenomenon of interest. Compare with *longitudinal study.*

cross-transfer the transfer of *biological material* from one medium to another.

crown court in England and Wales, the higher court of first instance for criminal cases.

crown prosecutor the prosecutor for the state, particularly for Commonwealth nations and possessions.

cruel and unusual punishment punishment that exceeds reasonable standards of severity. Cruel and unusual punishment is prohibited by the *Eighth Amendment* of the Constitution of the United States. An example of a punishment which might be considered cruel and unusual is a botched *electrocution* during which the condemned catches on fire and has to have multiple administrations of electrical current before death occurs. Confinement in a *supermax prison* has also been mentioned as cruel and unusual punishment.

cruelty to animals the gross neglect, abuse, or killing of domesticated animals. One of the three alleged indicators of a future homicidal *offender*, the other two are *bedwetting,* and *fire setting.*

crystal meth the crystalline form of the drug *methamphetamine.* Crystal meth gained popularity in the United States in the 1980s and 1990s.

CSA abbreviation for *Covenant, Sword, and Arm of the Lord.*

C.S.I. abbreviation for *crime scene investigation* and *crime scene investigator.*

CSI effect   a potential bias experienced by jurors and other participants in the criminal justice process as a result of exposure to movies and television programs where *crime scene investigation* is featured.

culpable   legally blameworthy for a behavior in question.

culprit   an individual suspected of committing a crime. See *suspect*.

cultural criminology   a branch of criminology that focuses on the importance of culture and context in understanding the genesis of crime and society's response to it.

culture conflict   the termed coined by 20th-century sociologist and criminologist *Thorsten Sellin* in his monograph Culture Conflict and Crime to describe the clash of norms for those who emigrated to the United States from abroad. See *primary conflict, secondary conflict*.

culture of terror   term used to describe the violent world of underground drug trafficking in large cities.

cumulative disadvantage   the net effect of poverty, *social disorganization*, physical, sexual, or emotional abuse or other problems associated with inner-city urban life on those who live there.

Cunanan, Andrew (1969–1997)   the individual who engaged in a *crime spree* that included the murder of famous designer Gianni Versace and several others. Cunanan eventually took his own life.

curfew   a time designated by parents or officials beyond which minors are not permitted to be out unsupervised.

custodial sentence   a sentence in a criminal case where the convicted offender is subjected to some form of custody, such as prison or secured drug treatment.

Customs and Border Patrol, United States   a division of the *Department of Homeland Security* responsible for preventing the entry of illegal aliens and smugglers into the United States. Customs and Border Patrol employs a variety of means to detect the passage of people and vehicles, including electronic surveillance and air, marine, and horse patrols. In addition to apprehending illegal aliens, Customs and Border Patrol routinely seizes substantial amounts of illegal drugs being smuggled into the United States.

cutpurse   a thief who steals by cutting purses. Compare with *pickpocket*.

cyanoacrylate   a substance used to make visible *latent prints* on nonporous surfaces.

cyber attack   an attack leveled at computer information systems.

cyberbullying *bullying* facilitated by e-mail, social networking, or the electronic posting of mean-spirited comments and messages about a person (such as a student), often done anonymously.

cyber crime crime committed with, or facilitated by, computers, the Internet and other on-line services. See *computer virus, hacker*.

cyber grooming the practice of eliciting the trust of potential victims by sexual predators using the Internet.

cyber jihad the use of the Internet by Islamic extremists to spread their messages of terror.

cybersecurity measures taken to prevent computer systems from being compromised by internal or external threats.

cyber stalking the *stalking* of individuals using a computer or other device that permits access to the Internet.

cyber surveillance proactive efforts undertaken to identify potential threats using computers and the Internet.

cyber threat threat posed by efforts to damage or disrupt a computer network.

cycle of violence the exposure of children to violence by their families which in turn increases the likelihood that they too will engage in violent behavior toward others, including their children, when they grow older. Compare with *intergenerational transmission of crime*.

C

dactylography   the study of using fingerprints for identification. See *fingerprint*.

Dahmer, Jeffrey (1959–1994)   a notorious *serial killer* who murdered a number of young gay men, primarily in Milwaukee, Wisconsin. Dahmer set himself apart from most other serial killers by torturing, sexually assaulting and dismembering his victims, cannibalizing, and then refrigerating some of their remains. Convicted of 16 counts of murder in 1991 and sentenced to life in prison, he was later murdered by another inmate. See *abuse of a corpse, necrophagia, serial murder.*

daisy chain scam   an illegal operation where companies create a chain of affiliates, each selling products to another, in order to manipulate the products' prices. See *corporate crime.*

Dalkon Shield   an intrauterine device (IUD) that was widely distributed and marketed by the A. H. Robins Company despite the fact it was responsible for the deaths of a number of women. See *corporate crime.*

DARE   abbreviation for *Drug Abuse Resistance Education.*

dark figure of crime   same as *hidden crime*. See *reported crime, victimization survey.*

Dark Triad   the combination of three personality traits—*psychopathy*, Machiavellianism, and *narcissism*. Individuals possessing the Dark Triad are considered at higher risk of becoming involved in aggressive, impulsive and criminal behavior.

data mining   the statistical exploration of databases in order to answer specific questions or uncover hidden patterns, including those related to crime.

date rape   the unlawful sexual violation of an individual during the course of a date. The rapist may use any of a variety of methods to get the victim to submit, including force, threats, or drugs like *Rohypnol*. Compare with *acquaintance rape.*

date rape drug   a drug like *Rohypnol* used to sedate one's date in order to facilitate sexual conduct without resistance. See *acquaintance rape, date rape, dating violence.*

D

dating violence   aggressive or assaultive behavior that occurs in romantic relationships.

Daubert rule   a standard providing a rule of evidence regarding the admissibility of expert testimony introduced in court. Derived from *Daubert v. Merrell Dow Pharmaceuticals.* See *junk science.*

Davis, Richard Allen (1954–present)   an ex-convict who kidnapped and murdered *Polly Klaas* in 1993. Davis sits on *death row.*

day fine   a financial punishment where the offender pays a portion of daily income. The amount paid also depends on factors, such as aggravating or mitigating circumstances, and the seriousness of the offense. Day fines, which originated in Scandinavian countries, have been shown to not jeopardize public safety.

day reporting center   a correctional program that permits convicted offenders to live at home, but receive services at a center. Day reporting centers can take a variety of forms and can be combined with other correctional options, such as electronic monitoring and intensive supervision probation. The primary advantage of day reporting centers is their cost savings over prison, halfway houses, and other residential correctional options.

day treatment   a rehabilitative alternative where the convicted offender undergoes correctional treatment during the day, usually at a day treatment center, and returns home at night.

deadly force   force used by law enforcement officers that can result in death. Concern over the use of deadly force has spawned interest in the development of *non-lethal weapons,* such as the *stun gun.*

dead man walking   an expression that conveys a prison inmate condemned to death on the way to execution. The phrase became popular with the publication of the nonfiction book *Dead Man Walking* by Sister Helen Prejean and the subsequent motion picture of the same name. See *death penalty, death row.*

death bed confession   revelations of criminal responsibility by one who is about to die. See *dying declaration.*

death camp   a camp designed for the extermination of its inmates. Death camps were used by the Nazis during the *Holocaust.*

death certificate   an official document issued by a *medical examiner* or *coroner* and signed by a physician attesting to the nature and cause of a person's death. Death certificates typically include not only identifying information, such as age, sex, and race, but also the *cause of death* and *manner of death.* See *coroner.*

death eligible case a criminal case where the *death penalty* is a possible punishment.

death investigation an official investigation into the *manner of death* and *cause of death*.

death of a thousand cuts a form of execution in which the individual sustains multiple superficial cuts which bring about death only after hours and sometimes days of slow bleeding.

death penalty punishment for a crime that results in the execution of the *defendant*.

Death Penalty Information Center a nonprofit organization that analyzes and disseminates information related to *capital punishment*.

death row the cells where prisoners sentenced to death spend their time awaiting *execution*. See *dead man walking*.

death row inmate a convicted *offender* awaiting *execution* on *death row*.

debtor's prison a penal facility used in the past to punish those who could not meet their financial obligations.

debt penalty term that conveys the financial burden incurred by those who are processed in the criminal justice system.

death squad a group of individuals whose purpose is to carry out *summary executions*.

death warrant see *execution warrant*.

decarceration the movement or practice of reducing offender populations in correctional facilities.

deceased another term for dead. One who is dead.

decomposition the natural breakdown of organic material on cessation of life. See *adipocere, body farm, putrefaction*.

decriminalization the removal or reduction of criminal penalties for acts previously defined in the law as illegal. Perhaps the best-known example is the movement to decriminalize *marijuana*. See *National Organization for the Reform of Marijuana Laws (NORML)*.

de facto legal term meaning regarding facts. Compare with *de jure*.

defeminization the removal of a woman's breast or other female organs during an assault or homicide. Such behavior, which is typical of *disorganized offenders* usually occurs postmortem. See *abuse of a corpse, depersonalization, serial murder*.

defendant   a person facing criminal charges. Compare with *codefendant*.

defense attorney   a lawyer whose responsibility is to ensure that the rights of the accused are upheld. See *appointed counsel, public defender*.

defense witness   a witness whose *testimony* is believed to serve the interests of a defendant's criminal case. Compare with *prosecution witness*.

defense wounds   cuts or other injuries sustained by a victim in attempting to ward off a physical attack involving a weapon. Defense wounds most often are found on the hands and arms.

defensible space   term coined by architect Oscar Newman in a book by the same name, defensible space conveys the immediate environment surrounding an individual that should be free of the threat of crime and other risk factors. See *Crime Prevention Through Environmental Design (CPTED)*.

deferred prosecution   a prosecution that is postponed on the condition that the defendant comply with specified conditions.

definite sentence   a *sentence* consisting of a fixed term of confinement. Also referred to as a fixed sentence. Compare with *indefinite sentence*.

degraded evidence   *evidence* that has lost some or all of its analyzability due to time or environmental conditions.

deinstitutionalization of status offenders (DSO)   the movement to keep juvenile status offenders from being held in secure confinement. No minor accused of an act which would not be criminal if committed by an adult may be securely detained in a jail, lockup, or juvenile detention center. DSO is predicated on the notion that secure confinement is harmful to those who have not engaged in delinquency.

de jure   legal expression meaning regarding the law. Compare with *de facto*.

deliberate indifference   the intentional disregard of the consequences of an act of commission or omission.

delinquency   acts by a juvenile which, if committed by an adult, would be treated as criminal. Compare with *status offense*.

delinquent   one who engages in *delinquency*. Also, one who is adjudicated as a delinquent by a juvenile court.

delinquent subculture   a subset of youths in society who are set apart by a set of beliefs, values, and activities that are vastly different from conventional society. See *subculture theory*.

demand reduction   efforts intended to reduce potential or existing desires of substance abusers to take drugs. Demand reduction depends on educating practicing drug users or those at risk about the legal, health, financial, and other consequences of *substance abuse*. Demand reduction can also encompass treatment. Compare with *supply reduction*.

demography the field that analyzes the characteristics and movements of human populations. Demographic factors, such as mobility, migration, and crime influence one another.

demonology the study of demons and other non-human spirits. See *goths*, *Satanism*.

demonstration project a pilot program designed to demonstrate the efficacy of a new approach to the prevention or control of crime or delinquency, or to the improvement of the administration of justice.

demonstrative evidence evidence which takes the form of an object. Refers to items shown in court, such as physical objects, graphs, pictures, blow-ups of documents, models, and other devices which are intended to clarify the facts.

de novo a legal term meaning fresh or new, indicating that the court hearing a case need not consider the findings of earlier decisions.

dental stone a substance used at crime scenes to make castings of tire and shoe impressions. Dental stone works better than plaster of Paris because it is more resistant to breakage. See *moulage*.

dependent variable in explanatory research, the phenomenon to be explained. Compare with *independent variable*.

depersonalization efforts made by a killer to cover or otherwise obscure the identity of a victim. See *disorganized offender*.

deportation the physical expulsion of an individual to his or her country of origin.

deposition out-of-court testimony of a witness to be used later in court proceedings.

depraved mentally sick. See *depraved heart murder*.

depraved heart murder extremely negligent or atrocious behavior that results in the death of another. An example of depraved heart murder would be a man who wrests a life jacket from a child to save his own life, thereby leaving the child to drown.

depraved indifference murder see *depraved heart murder*.

Depravity Scale, The a scale developed in order to standardize the depravity of very serious crimes. The Depravity Scale focuses on the nature of the act as opposed to the characteristics of the person who performed the act. Compare with *Gradations of Evil Scale*.

D

deprivation model   a model advanced by sociologist Gresham Sykes to explain how inmates cope with the deprivations inherent in prison life. Compare with *importation model.*

deputy   an officer subordinate to a *sheriff.*

De Salvo, Albert (1930–1973)   a *serial killer* who is alleged to have murdered at least 13 women in the Boston area from 1962 to 1964. Despite DeSalvo's confession to 11 of the murders, there are many who do not believe he was the true *Boston Strangler.* DeSalvo, who was defended by attorney *F. Lee Bailey*, eventually was killed by another inmate in prison. See *serial murder.*

design capacity   the capacity of a prison or other correctional facility as intended by the designer. The design capacity, intentional to ensure the safety and security of both inmates and staff, is often exceeded due to *overcrowding.* Compare with *rated capacity.*

designer drugs   trendy substances of abuse, often analogs of a drug like fentanyl. Popular designer drugs are responsible for many emergency room admissions since users often are unaware of the drugs' effects and hazards. See *club drugs.*

desistance   the refraining from criminal behavior. Criminologists study desistance in order to determine why offenders discontinue engaging in illegal behavior. See *criminal career.*

detainee   one who has been detained or held in custody by law enforcement or other authorities. Compare with *arrestee.*

detective   a law enforcement officer, usually working in plain clothes, who specializes in investigating unsolved felonies, such as homicides, sexual assaults, robbery and burglary, as well narcotics and *vices.* Compare with *private detective.*

detention center   a facility for the temporary or short-term confinement of accused or adjudicated youths.

detention home   same as *detention center.*

determinate sentence   a fixed amount of time a convicted offender must serve in prison. The length of a determinate sentence is prescribed by law and cannot be modified by the sentencing judge or by correctional officials. Same as fixed sentence. Compare with *indeterminate sentence.*

determinism   the theory that human conduct is determined by biological, psychological, environmental or other causes. Implicit in determinism is that offenders are less culpable for their actions. See *biological determinism.*

deterrence   one of the major justifications for punishment, deterrence conveys a legal threat to prospective offenders that criminal sanctions purportedly pose. Deterrence is thought by many to dissuade would-be offenders from engaging in criminal behavior. See *general deterrence, specific deterrence*. Compare with *incapacitation, rehabilitation, retribution*.

detoxification center   a treatment facility designed to permit those who are addicted to alcohol or other drugs to safely withdraw from the substance of abuse. Also referred to as a detox center.

developmental criminology   the branch of criminology interested in the psychological and social development of offenders throughout the life course.

developmental trajectory   the expression of delinquent traits at various points along the life course. Also referred to as developmental pathway.

deviance   acts or behaviors that deviate from what is considered normal or appropriate.

Devil's Island   a former settlement and *penal colony* in French Guiana. Devil's Island was in use from 1852 to 1946. Due to the unhealthy climate, many prisoners died and few managed to escape from Devil's Island. See *penal colony*.

Diagnostic and Statistical Manual of Mental Disorders (DSM)   a compilation published by the American Psychiatric Association that inventories all known mental disorders. The DSM has had several editions, the most recent of which is the DSM-V (2013).

differential association   a criminological theory put forth by the late *Edwin H. Sutherland* that posits that persons learn to become criminals through their close, intense association with others disposed to criminal behavior.

differential association reinforcement theory   a restatement of *differential association* in light of behaviorism. Differential association reinforcement theory, developed by Robert L. Burgess and Ronald Akers, emphasizes the role of operant conditioning in the way crime is learned.

digital abuse   the use of computers and other devices to harass, intimidate, or humiliate others. Compare with *cyberbullying*.

digital crime   see *computer crime*.

digital evidence   data stored in digital form that has the potential of being used in a criminal case.

digital forensics   a specialty within forensic science that focuses on the analysis of digital data on computers and other devices, such as smartphones.

digital harassment   see *digital abuse*.

Dillinger, John (1903–1934)   a notorious bank robber of the 1930s. Federal agents led by FBI agent *Melvin Purvis* shot and killed Dillinger outside the Biograph Theater in Chicago.

diminished capacity   a defense used by accused offenders which asserts that they could not control their behavior because of a mental or other condition that diminishes their capacity to abstain from crime.

Dinitz, Simon (1926–2007)   a criminologist and long-time professor of sociology at The Ohio State University best known for his research in social psychiatry, dangerous offenders, and with *Walter C. Reckless*, containment theory.

diplomatic immunity   insulation from criminal prosecution extended to foreign members of diplomatic corps. Diplomatic immunity stirs controversy when persons granted immunity do not have to face criminal prosecution, even for very serious offenses.

direct examination   the first examination of a witness in a court proceeding. Compare with *cross examination*.

direct file waiver   a waiver that gives a prosecutor the discretion to decide whether to pursue a case in juvenile or adult criminal court.

dirty urine   a urine specimen containing prohibited drugs. Drug treatment programs require clients to submit urine specimens to monitor drug use. A dirty urine test finding can result in loss of privileges, expulsion from the treatment program, or in the case of those on *probation*, revocation. See *substance abuse*.

disarmament   the removal of arms from those who pose a threat to the safety and security of a community.

disaster preparedness   efforts designed to understand and anticipate the challenges associated with natural or manmade disasters.

disaster response team   a group of paid staff or volunteers that offers various forms of assistance in the wake of a disaster. Compare with *first responder*.

discharge   the release of an inmate from a prison, jail, or other form of confinement.

discharge residue   the residue left as a result of a firearm discharge. Discharge residue can include such chemicals as lead, antimony, and barium. Also called *gunshot residue*.

disclosure   the revelation of information about a person or entity.

discovery   the legal practice by both parties of exchanging information.

D

discretion   the ability that inheres in criminal justice roles to make subjective judgments that affect those processed in the criminal justice system. Discretion permits officials to circumvent legislative intent designed to treat offenders more harshly, such as *three strikes and you're out*.

discretionary justice   the ability of law enforcement officers, prosecutors, and other functionaries in the criminal justice system to use their own judgment on how or even whether to process offenders or cases. While discretion is inherent in much of the criminal justice process and perhaps even desirable, it can lead to disparities in the treatment of similarly situated offenders. Its abuse can result in gross miscarriages of justice.

discretionary release   the ability of correctional officials to release offenders when extraordinary circumstances like *overcrowding* dictate action.

discrimination   policies or actions which denigrate and abridge the rights of selected classes of people.

disembowel   to cut the abdomen in such as way as to permit the intestines to spill out or be exposed. Same as *eviscerate*.

dishonesty offence   in Australia, being intentionally dishonest by influencing a public official and causing a monetary loss to the Commonwealth.

disintegrative shaming   shaming which does not attempt to reintegrate the *offender* back into society. Disintegrative shaming results in *stigma* for the offender. See *integrative shaming*, *shaming penalties*.

dismember   to cut the limbs and head from the torso of a human body.

dismissal   the cancellation of a criminal case. Dismissals can occur for a variety of reasons.

dismissal with prejudice   the *dismissal* of a criminal case specifying that the *defendant* cannot be recharged with the crime in question. Compare with *dismissal without prejudice*.

dismissal without prejudice   the *dismissal* of a criminal case where it is possible for the defendant to be charged again with the crime in question. Compare with *dismissal with prejudice*.

disorderly conduct   a minor charge, generally a misdemeanor, used by police to charge drunken or otherwise publicly disturbing behavior.

disorganized offender   a *serial killer* who tends to be of lower intelligence, kills spontaneously rather than through careful planning, and engages in *depersonalization* of his victims. Compare with *organized offender*. See *serial murder*.

disparity   an inconsistency in the way offenders or their criminal cases are treated compared to similar cases. See *disproportionate minority confinement, sentence disparity.*

disposition   the conclusion of juvenile or criminal court proceedings, often with an *adjudication* in an adult case or the imposition of a sentence in a criminal case.

disproportionate minority confinement   the confinement in detention, jails, prisons, or other facilities of minorities (juveniles in particular) in percentages out of proportion to their representation in the general population. Beginning in the late 1980s, the *Office of Juvenile Justice and Delinquency Prevention* made disproportionate minority confinement one of its priority areas for research and policy development. Despite these efforts, disparities in processing remain. See *disparity.*

dispute resolution   see *alternative dispute resolution.*

distributive justice   *justice* focused on the fair distribution of outcomes. Compare with *procedural justice.*

district attorney   same as *prosecuting attorney.*

disturbing the peace   a generally minor offense, most often a misdemeanor, consisting of making too much noise in public.

diversion   the formal routing of offenders away from traditional criminal or juvenile justice processing for a specified period of time and under certain conditions, after which their charges will be dismissed. In theory, diversion minimizes the *stigma* associated with criminal conviction. In practice, the use of diversion sometimes leads to *net widening.*

DNA   deoxyribonucleic acid, the building blocks of humans and other living things, is a molecule that encodes genetic instructions. See *DNA testing.*

DNA testing   the analysis of DNA in criminal cases. DNA is unique to the individual, making it the perfect means of matching a particular offender to a crime. The analysis of DNA is used both to incriminate and exculpate those suspected and convicted of crimes. See *Innocence Project, wrongful conviction.*

docket   the totality of cases awaiting processing in a criminal court.

domestic burglary   the burglary of a residence. Compare with *commercial burglary.*

domestic murder   the killing of an intimate, most often a spouse or significant other. See *intimate partner violence.*

domestic terrorism   terrorist acts committed on domestic soil, often by citizens. Examples of domestic terrorism include the *Oklahoma City bombing* and arson committed by *ecoterrorists*. Compare with *international terrorism*. See *Nichols, Terry L.*, *McVeigh, Timothy*.

domestic violence physical or sexual assault of an intimate, most often a spouse, other relative, or significant other. See *domestic violence shelter*, *intimate partner violence*.

domestic violence court a court that specializes in adjudicating cases of *domestic violence*. See *specialty court*.

domestic violence shelter a house or other building maintained for the purpose of providing safe housing for women who have been physically or otherwise abused by a spouse or partner in a domestic setting, or for individuals equally affected by the violence.

doorstep crime a crime characterized by efforts to bilk residents by selling them bogus products or services or trying to enter the home using a false identity.

double bunking the practice of housing two jail or prison inmates in a cell intended for one. Also referred to as double celling. See *design capacity*, *overcrowding*, *rated capacity*.

double deviance term used to convey the harsher view of female offenders whose behavior does not conform to social expectations.

double jeopardy a situation where a criminal defendant faces trial for a crime after an acquittal or conviction for the same crime.

downers slang term for central nervous system depressants and other drugs known to have a calming or tranquilizing effect. Compare with *uppers*.

dowry death the intentional murder or induced suicide of a newly married woman in India due to the perceived insufficiency of her dowry.

Dream Team term used to describe the group of high-profile criminal defense attorneys that defended *O. J. Simpson* against the charges that he murdered Nicole Brown Simpson and Ronald Goldman. The Dream Team included at one time or another *F. Lee Bailey*, *Johnnie Cochran*, Harvard professor Alan Dershowitz and Robert Shapiro. Ultimately, they were successful in obtaining an acquittal for their client.

drift according to David Matza in his book *Delinquency and Drift*, the process by which youths in a delinquent subculture move back and forth between conventional activities and illegal activities.

drive-by shooting a shooting from a moving motor vehicle. Drive-by shootings, most closely identified with conflicts between gangs, occasionally result in the injury or death of *innocent bystanders*.

D

driving under the influence (DUI)   the act of operating a motor vehicle under the influence of alcohol or other drugs. Also *driving while intoxicated*. See *Mothers Against Drunk Driving (MADD)*.

driving while Black   a derogatory expression used to describe the practice by law enforcement officers of pulling over Black motorists where there is no violation of the law. See *racial profiling*.

driving while intoxicated   same as *driving under the influence*.

drowning   death caused by immersion in a liquid, most often water.

drug   an organic or chemical compound, which may or may not have therapeutic effects, used for its mind- or consciousness-altering effects. Illegal drugs include *crack, marijuana,* and *methamphetamine.* See *substance abuse.*

drug abuse   the illegal use of substances of abuse. This is also the term for the offense which consists of engaging in this type of illegal use. Generally, drug abuse as an offense is a *misdemeanor* punishable by jail time, a fine, or both.

Drug Abuse Resistance Education (DARE)   a program where law enforcement officers instruct school students about the dangers of *substance abuse,* gang membership, and violence. Despite DARE's widespread popularity, its effectiveness in reducing substance abuse has been challenged by evaluation studies. In response, DARE has made changes to its curriculum.

Drug Abuse Warning Network (DAWN)   a large-scale data collection system sponsored by the *Substance Abuse and Mental Health Services Administration (SAMSHA).* Using data reported by hospital emergency departments and medical examiners, DAWN collects information on what drugs are being used, which ones are related to drug deaths, and which are currently in vogue among users. See *Drug Use Forecasting (DUF).*

drug cartel   a criminal organization whose activities include the cultivation, manufacture, distribution, and sale of illegal drugs. See *Cali Cartel, cartel, drug trafficking, Medellin cartel, Zetas.*

drug courier   one who transports narcotics or other drugs, often on their person. Compare with *mule.*

drug court   a *specialty court* designed to administer justice and meet the treatment needs of those suffering from *substance abuse.*

drug czar   a state or federal governmental official in charge of anti-drug policy. For the United States, the drug czar is the director of the *Office of National Drug Control Policy (ONDCP).*

Drug Enforcement Agency (DEA) a branch of the U.S. Department of Justice responsible for investigating the violation of federal controlled substance laws.

DRUGFIRE Program refers to the computer technology that permits law enforcement officials to link firearms evidence in shooting investigations. The DRUGFIRE Program was developed by the *Federal Bureau of Investigation (FBI)*. See *linkage blindness*.

drug identification the process of determining the type of drug in a sample.

drug interdiction efforts directed at interrupting the flow of illegal drugs, particularly those being brought into a jurisdiction from the outside. See *Drug Enforcement Administration (DEA)*.

drug kingpin the head of a criminal organization whose principal activity is the distribution of illegal drugs. Compare with *boss*.

drug legalization the movement to make the cultivation, possession, and use of drugs legal. See *National Organization for the Reform of Marijuana Laws (NORML)*. See *decriminalization*.

drug overdose the introduction into the human body of a drug in an amount in excess of a normal dose with often toxic or lethal results. See *toxicology*.

drug paraphernalia material or devices used to make or take drugs. Examples include syringes, needles, spoons, and pipes.

drug possession the crime of having small amounts of drugs under one's control.

drug testing the process of taking and analyzing human blood, urine, hair, or other substances in order to determine the presence and type of drugs used.

drug trafficking the selling and distribution of narcotics or other drugs for the purpose of making a profit. Penalties for drug trafficking are much greater than those for possession and use.

Drug Use Forecasting (DUF) a program in the United States designed to collect data on jail inmates regarding the nature and extent of their drug use. See *substance abuse*.

drunk driving the act of operating a motor vehicle while under the influence of alcohol. See *driving under the influence, Mothers Against Drunk Driving (MADD)*.

drunkenness a minor criminal charge used against those whose intoxication results in a public disturbance or nuisance. See *public intoxication*.

DSM-III-R abbreviation for *Diagnostic and Statistical Manual*, Third Edition, Revised.

D

DSM-IV abbreviation for *Diagnostic and Statistical Manual*, Fourth Edition.

DSM-V abbreviation for *Diagnostic and Statistical Manual*, Fifth Edition.

dual diagnosis having two problems in need of treatment at the same time. A mentally ill person with a *substance abuse* problem is an example of someone with a dual diagnosis.

dualistic fallacy the mistaken notion that criminal populations under study are distinct from the general population which is assumed to be composed of non-criminals. The dualistic fallacy is exemplified by the once held belief that criminals came solely from the lower socioeconomic classes. The existence of *upperworld crime* disproves that belief.

dueling the now outdated practice of settling disputes between two individuals by fighting with guns, swords, knives, or other weapons, often to the death. One of the most famous duels in American history was that between Alexander Hamilton and Aaron Burr, where the former was mortally wounded. Dueling is virtually obsolete in Western societies, having been banned for nearly two centuries.

due process the notion in Anglo-American law that those accused of a crime have certain rights to fair and impartial processing, including the presumption of innocence.

DUI abbreviation for driving under the influence of drugs or alcohol. Same as *OMVI* or operating a motor vehicle while intoxicated. See *drunk driving*.

dump site the place where a homicide victim's body is deposited which may be different from the *crime scene*.

dungeon an unpleasant, often subterranean chamber used for the confinement, torture, or execution of prisoners.

dunking stool used in early America, a stool fixed to the end of a pole resting on a fulcrum to punish wrongdoers. The individual seated on the dunking stool was submerged in water in order to elicit a confession or punish them for wrongdoing. See *torture, trial by ordeal*.

duress threat or force used illegally to make someone do something against his or her will. Duress is used to mitigate the seriousness of a crime.

duress defense a criminal defense based on the premise that the individual was under *duress* at the time of the offense.

Durham rule a 1954 standard that a person charged with a crime is not responsible if the act was the result of mental disease or mental defect. Compare with *M'Naughten rule*.

Durkheim, Emile (1858–1917) an early French sociologist who introduced the concept of *anomie*. Durkheim, who analyzed French suicide and other statistics for patterns and regularities, argued that crime serves a positive function in society. His work had a profound influence on a number of sociologists and criminologists, including *Robert K. Merton*.

dusting the use of a soft bristled brush to apply powder to suspected *latent prints*. See *lifting*.

DWI abbreviation for driving while intoxicated. Same as *DUI*.

dying declaration an oral or written statement made by one who is about to die that may include accusatory or exculpatory information. Dying declarations generally are regarded as truthful since a person who is about to die has little incentive to lie. See *deathbed confession*.

dysfunctional family a family characterized by marital discord, physical, sexual, mental, or emotional abuse or issues, neglect, or other problems that prevent the family's normal social functioning.

Eastern State Penitentiary a prison in Pennsylvania intended to provide both incarceration and labor. Opened in 1929, Eastern State Penitentiary became the model for countless other prisons around the United States and the world. It was abandoned in 1971. See *Pennsylvania system*.

echo boom a population boom created by the offspring of a *baby boom*. Echo booms result in larger numbers of persons of offending ages in the population. And just like a baby boom, an echo boom also strains the resources of the criminal justice system.

ecological criminology see *Chicago School of Criminology*.

ecological fallacy the mistaken tendency to draw inferences about individuals based on aggregate data. An example is to assume that a poor person is a criminal because of the statistical relationship between social disadvantage and crime.

ecological school of criminology same as *Chicago School of Criminology*.

ecology of crime the geographical and social distribution of crime, as well as the meaning of such distributions.

economic crime any crime that involves financial assets.

economic disadvantage the lack of jobs or other means of financial support. Economic disadvantage is thought to contribute to crime and social disorder.

economic espionage activities related to the theft of trade secrets for commercial purposes instead of purely national security.

ecoterrorism violent crimes committed in the name of preserving the environment and other natural resources.

ecoterrorist a person who engages in *ecoterrorism*.

Ecstasy an illegal drug that reduces anxiety and increases euphoria. Also known as MDMA. See *substance abuse*.

ectomorph a body type characterized by thinness or slightness of build. Compare with *endomorph, mesomorph*. See *Sheldon, William*.

edgework   predominantly masculine, risky pursuits of individuals that are related to offending.

effect size   a descriptive statistic that conveys the strength of a research finding.

Eighth Amendment   the amendment of the U.S. Constitution that prohibits *cruel and unusual punishment.*

elder abuse   the physical, mental, or emotional abuse of an older person, most often a parent or grandparent by a caregiver. Elder abuse is occasionally a problem in nursing homes.

elder court   a *specialty court* designed to meet the needs of senior citizens with special needs.

election fraud   whether occurring before, during, or after an election, illegal efforts intended to alter the outcome of the election.

electrical equipment conspiracy   a major price-fixing conspiracy in the early 1960s by electrical equipment manufacturers, including General Electric and Westinghouse, which resulted in heavy fines and jail terms for some of those involved. This conspiracy was one of the first major cases where corporations were held criminally responsible for the actions of their executives. See *corporate crime.*

electric chair   a chair designed for the *execution* of condemned prisoners by the application of high voltage electricity. The electric chair has been known to result in botched executions, prompting critics to label it *cruel and unusual punishment.* Many states that employed the electric chair have changed their means of execution to *lethal injection.* Compare with *gas chamber, hanging.* See *death penalty.*

electrocution   a form of *execution* in which the *electric chair* is used.

electronic crime   crime involving electronic financial payment systems.

electronic identity theft   identity theft facilitated by the use of electronic devices.

electronic monitoring   the use of an electronic device that permits authorities to limit and monitor the mobility of an *offender* placed in home detention. Electronic monitoring is used for both accused and convicted misdemeanants and felons. Such devices permit the control of a large number of offenders by a relatively small number of staff members. The technology has not yet been perfected, which can result in false readings by the electronic equipment. Electronic monitoring is referred to as tagging in the United Kingdom. Also referred to as ELMO.

electrophoresis   see *gel electrophoresis.*

electrostatic detection apparatus a device used to lift images from paper without altering the paper in any way, making the sample available for subsequent analysis.

elimination prints fingerprints taken from those who may have been in or around a crime scene in order to distinguish the prints of innocent persons from those of possible criminal suspects.

Elmira Reformatory established in 1876, a New York correctional institution that became the model of reformatories for young offenders. What made Elmira Reformatory distinct from earlier reformatories was its emphasis on education in the trades. With the advent of Elmira came the *indeterminate sentence, parole,* and attempts at *inmate classification.*

embezzlement the surreptitious theft of money by a person in a position of trust. Examples include theft by an investment counselor or bank teller. See *white-collar crime.*

emotional abuse abuse of another caused by humiliation, verbal aggression, intimidation, or other nonphysical efforts to denigrate. Compare with *physical abuse.*

employee crime crime committed against a business or other organization by a person in its employ.

employee theft the unauthorized taking of goods or services from a business or organization by those in its employ. Research shows that a great deal of employee theft stems from the perception of workers that they need to steal in order to restore a sense of equity between themselves and their employers. See *employee crime.*

endomorph a body type characterized by excessive weight in proportion to height. Compare with *ectomorph, mesomorph.* See *William Sheldon.*

endophenotype a hereditary characteristic that is normally associated with some condition, but is not a direct symptom of that condition. Some endophenotypes include risk taking and narcissistic traits. See *biocriminology.*

enhanced interrogation techniques techniques used during the interrogation of a criminal suspect which bring about fear, discomfort, or pain. Enhanced interrogation techniques were used against some prisoners at *Guantanamo Bay.* See *torture, waterboarding.*

Enron scandal the bankruptcy and collapse of the Enron Corporation as a result of creative accounting procedures which inflated the company's value. See *corporate crime, white-collar crime.*

entrance wound a wound in human tissue created when penetrated by a bullet or other projectile. See *bullet track.* Compare with *exit wound.*

E

entrapment   the practice of law enforcement of enticing a person to commit a crime who otherwise would not consider engaging in criminal behavior.

environmental crime   the destruction or contamination of the environment through neglect or purposeful action. An example of environmental crime is the poor disposal practices that render land unusable. Compare with *ecoterrorism*.

environmental criminology   a branch of criminology that focuses on aspects of the environment and their impact on people. Compare with *green criminology*.

environmental hazard   any substance or condition that poses a threat to the environment.

epidemic   the widespread contagion of a disease or other socially harmful phenomenon, such as crime, in a specific area within a specific period of time. See *tipping point*.

epidemiology   the science of determining the incidence, prevalence, and causes of diseases and other problems affecting human populations.

Equity Funding scandal   a corporate swindle where the corporation created thousands of phony insurance policies. Consequently, both reinsurers and stockholders lost millions of dollars. The scandal resulted in indictments and prison terms. See *corporate crime, white-collar crime*.

escalation   an increase in the amount of crime an offender commits. Criminologists study escalation to determine the age or point at which escalation occurs. Compare with *desistance*.

escape   the unauthorized or illegal flight from custody of an accused or convicted offender.

escape artist   one adept at escaping from secure devices, such as handcuffs and straightjackets, or places, such as jails or prisons.

escape from lawful imprisonment   see *escape*.

escape risk   an individual considered likely to attempt an escape from confinement.

espionage   the practice of an agent or spy surreptitiously gathering information, usually for political or military uses, that is not intended for their dissemination. See *industrial espionage, Walker spy ring*.

ethics   the science of appropriate human conduct.

ethnic cleansing   the policy of some countries or factions to systematically purge or annihilate members of certain races, ethnic groups, cultures, religious sects, or national origins. An example is violence between the Serbs and Croats in Bosnia during the 1990s. See *genocide*.

ethnic profiling   the use of racial, ethnic, national origin, religious, or appearance characteristics to identify individuals or groups for discriminatory treatment.

ethnic succession theory   the theory that the control of *organized crime* passes from one ethnic group to another over time.

ethnography   the systematic description of cultures and social phenomena, as the result of close observation and interaction.

etiology   the study of causation. In criminology, etiology focuses on the causes of crime, including individual, situational, environmental, and societal factors.

eugenic criminology   a theoretical perspective and its related policy which suggests that eliminating certain undesirable characteristics from the human population would reduce, if not eliminate, crime. Eugenic criminology found partial expression in the extermination policies of the Nazis in the 1930s and 40s. See *ethnic cleansing, Holocaust*.

European Court of Human Rights   an international court that serves the member states of the Council of Europe.

European Society of Criminology (ESC)   a society in Europe whose mission is to bring together those who work in the field of criminology. The ESC, which is headquartered in Lausanne, Switzerland, holds an annual conference and publishes the *European Journal of Criminology*.

Europol   the law enforcement agency of the European Union whose focus is international crime and terrorism. Compare with *Interpol*.

euthanasia   same as *mercy killing*.

event history analysis   a set of statistical methods used to analyze change of a phenomenon from one state to another. Event history analysis permits the analysis of longitudinal data even where all cases under analysis have not experienced the event of interest, such as involvement in delinquency.

Evers, Medgar (1925–1963)   a *civil rights* activist of the early 1960s who was murdered. See *hate crime, racist*.

eve-teasing   inappropriate sexual remarks or actions toward a woman in public.

evidence   physical material which can be used to establish a case for or against a *defendant* in a criminal case.

evidence-based practice (EBP)   criminal justice programs whose validity has been established through rigorous empirical studies. See *Campbell Collaboration, Cochrane Collaboration, systematic review.*

evidence tampering   the alteration, concealment, theft, or destruction of *evidence* in order to interfere with a criminal case.

evidence technician   an individual who assists law enforcement officers in collecting and analyzing *evidence* in criminal cases. Evidence technicians are often civilian employees with special training and experience in processing evidence. Compare with *criminalist.*

evidence trail   a series of linkages connecting evidence of a crime to a suspect. Compare with *audit trail.* See *linkage blindness.*

evidence vacuum   a small and powerful vacuum cleaner used by crime laboratory staff to collect particulate *evidence* of potential forensic interest.

eviscerate   to remove the intestines of a victim. Some serial killers eviscerate their victims, usually postmortem, in order to depersonalize them. See *depersonalization, disorganized offender.*

excessive force   more force than is necessary to subdue a suspect. See *chokehold, police brutality.*

excited delirium   bizarre aggressive behavior usually, but not always, drug-induced. In some cases, excited delirium can result in death.

exclusionary rule   the legal principle that prohibits the use of evidence illegally obtained by police. In some circumstances, the questionably obtained evidence is used, but the officer who illegally obtained it can be punished.

ex-con   short for ex-convict, one who has served time in prison.

exculpatory evidence   *evidence* which serves to prove that an individual did not commit a crime of which he or she has been accused. Compare with *inculpatory evidence.*

execution   the act of carrying out an official order to put a convicted offender to death. See *capital punishment, death penalty.*

executioner   one who puts condemned prisoners to death.

execution warrant   a warrant that authorizes the execution of an individual condemned to death.

exemplar   a legally admissible handwriting sample to use as a typical or standard specimen for comparison with a questioned piece of writing. See *questioned documents examination.*

exemplary project a project deemed so promising or successful that it should be recognized as worthy of replication by other jurisdictions. The former National Institute of Law Enforcement and Criminal Justice (NILECJ) of the 1960s and 1970s designated a number of innovative criminal justice practices as exemplary projects.

exhibitionism the practice of displaying one's genitals to others, most often for sexual gratification. Compare with *indecent exposure*.

exhumation the process of disinterring human remains in order to conduct toxicological tests or other necessary investigative procedures to confirm the circumstances surrounding the person's death. Exhumation may involve permission of the deceased's relatives, but in some cases may be ordered by a court.

exit wound the wound in human tissue made by a bullet or other projectile when leaving the body. The characteristics of an exit wound often differ from its corresponding entrance wound depending on the caliber and type of the projectile. Compare with *entrance wound, wound track*. See *wound morphology*.

ex-offender an individual who has been convicted of a crime, usually one who has served time in prison.

experimental criminology criminological research that employs random assignment in field experiments to test criminological theories or evaluate criminal justice programs. See *Academy of Experimental Criminology (AEC), experimental group*.

experimental group in an experiment, the group who receives the treatment. An example of an experimental group in a criminal justice evaluation would consist of those who participated in diversion or some other intervention. Compare with *control group*. See *Academy of Experimental Criminology*.

expert testimony testimony offered for either the prosecution or defense by a specialist whose expertise bears on the issue in question. See *Frye rule*.

expert witness one who provides *expert testimony* in a court of law. Expert witnesses can testify for either the *prosecution* or the *defense attorney*. See *expert testimony, Frye rule*.

exploratory research preliminary research on a topic on which little or nothing is known. Those conducting exploratory research frequently use qualitative methods or data from secondary sources.

explosive device a device used as a weapon that provides a sudden violent release of energy. Examples of explosive devices include bombs, grenades, and *improvised explosive device (IED)*. See *bomb squad, Boston Marathon bombing.*

explosive residue   remnants of chemicals left behind after an explosion. The analysis of explosive residue can tell officials not only the composition of the explosive device, but often also the manufacturer of the elements, which can in turn lead to the identification of those responsible for the bombing.

exposure to violence   the experience of witnessing or being in proximity to acts of violence. Exposure to violence can result in a variety of problems, including *post-traumatic stress disorder (PTSD)* and conduct problems, including aggressiveness. See *children who witness violence.*

expressive crime   a crime committed as a result of emotional excitement or distress rather than for personal gain. Compare with *instrumental crime.*

expulsion order   a legal document ordering a person's expulsion or summary removal.

expungement   the destruction or sealing of records of adjudication or conviction. In many jurisdictions, the records of juveniles are automatically expunged after a certain period of time. In adult courts, those with a single conviction may apply for expungement after a specified period of time.

exsanguination   loss or draining of blood from a body.

extended jurisdiction   the policy and practice of maintaining correctional jurisdiction over juveniles after they have reached the age of majority, usually 18 years of age.

extenuating circumstance   a circumstance surrounding the commission of a crime that somehow lessens its seriousness.

externalizing behavior   behavior that takes manifest forms, such as *aggression* toward others. Compare with *internalizing behavior.*

extortion   the illegal act of coercing payment by force or threat. See *protection racket.*

extradition   the legal transfer of an accused from one jurisdiction to the jurisdiction where the accused is wanted. An accused may fight extradition.

extrajudicial killing   a deliberated killing of a person not authorized by a previous judgment, therefore without due process or judicial involvement. Compare with *summary execution.* See *death squad.*

extralegal factors   factors unrelated to a criminal or a criminal case, but which may influence the disposition of the case. Examples of extralegal factors include age, race, ethnicity, and socioeconomic status, which have nothing to do with the crime. Criminologists have analyzed the influence of extralegal factors on criminal case dispositions. Also referred to as extralegal variables.

extra Y chromosome   a dated and now disproved theory in criminology that males with an additional Y chromosome instead of the more usual XY configuration are more prone to engage in criminal behavior. Research on the extra Y chromosome has been criticized for relying on confined populations that are not necessarily representative of the general population or of offenders in general.

eyewitness   a person who visually witnesses a crime. See *witness*.

eyewitness error   a mistake in identification made by an *eyewitness* of a crime.

eyewitness identification   the identification of a criminal suspect by an individual who actually made visual contact.

eyewitness testimony   *testimony* offered or given by an *eyewitness*. Research has revealed that eyewitness testimony is often unreliable, leading to the identification and sometimes the *wrongful conviction* of persons not responsible for the crimes in question.

facial composite a depiction of a suspect's face made by an artist or with the assistance of a manual or computerized system. See *composite, composite drawing, forensic art, Identikit.*

facial recognition system a computer-based system that employs video equipment to recognize a person's facial features from images on file.

facial reconstruction for purposes of identification, the process of taking a human skull and using clay, hair, or other materials to restore it as closely as possible to a likeness of the deceased. Computer technology has also been employed to reconstruct human likenesses. See *forensic art.*

failure to appear the charge leveled against those who do not show up for scheduled court hearings. Failure to appear often results in the issuance of a *warrant* for the person's arrest. See *bench warrant.*

fair trial a criminal trial characterized by *due process.*

faith-based programs crime or delinquency prevention or intervention programs promoted by or affiliated with the faith community.

Falcon and Snowman names given to Christopher Boyce and Andrew Dalton Lee, respectively, two spies who sold American military secrets to the Soviets in the 1970s. See *espionage, spy.*

FALN a radical Puerto Rican group operating between 1974 and 1983 and dedicated to achieving Puerto Rican independence from the United States. Their principal means of *terrorism* was bombings in both Puerto Rico and the United States.

false advertising commercial advertising which has the capacity to deceive the buying public. False advertising can also hurt competitors.

false arrest an arrest without *probable cause* or other authority of a person by law enforcement.

false confession a confession given to law enforcement officials by a person who did not commit the crime in question. Those who offer false confessions often desire public attention.

false impersonation   passing one's self off as another person. See *impersonating an officer*.

false imprisonment   the confinement of a person without regard for their rights. Sex offenders who kidnap and detain their victims could be charged with false imprisonment.

false positive   a predicted outcome that does not occur. In criminological prediction, an example of a false positive is a prediction that an offender will re-offend and they do not. Certain penal policies could result in the confinement of offenders who might never re-offend.

false pretenses   fraudulent representations made to obtain money, goods, or services.

familicide   *homicide* in which an individual kills a parent in addition to other family members, which can include siblings and/or grandparents.

family court   a court whose jurisdiction encompasses a wide range of family-related matters, including but not limited to delinquency, dependency, and other juvenile cases.

family group conference   a meeting of the offender, victim, their families and other supportive people to engage in a dialogue in an effort to bring about restoration and healing. Family group conferencing can take many forms and can be traced to New Zealand and Australia, but it has become popular in the United States. See *restorative justice, shaming penalties*.

family violence   violent acts taking place among those related or living together. Family violence can include both *physical abuse* and *psychological abuse*.

fantasy   often recurring mental imagery conjured up by many *sex offenders* and *serial killers* as a prelude to an offense. See *sex offender, serial killer*.

Faulds, Henry (1843–1930)   a Scottish physician and scientist credited with the development of *fingerprinting*.

FBI   abbreviation for the *Federal Bureau of Investigation*.

FBI Law Enforcement Bulletin   a monthly magazine published by the *Federal Bureau of Investigation* for law enforcement professionals. The *FBI Law Enforcement Bulletin* contains useful articles on how to investigate specific crimes and address legal issues confronting law enforcement.

fear of crime   the real or perceived extent to which citizens are concerned about their chances of criminal victimization.

Federal Bureau of Investigation (FBI)   the principal law enforcement and security arm of the U.S. Department of Justice. The Federal Bureau of

Investigation is responsible for addressing a wide array of violations of federal law. In addition, the FBI offers its investigatory services and crime laboratory to state and local authorities in certain circumstances. *Quantico, VA* is the site of its national academy.

Federal Emergency Management Agency (FEMA)   the federal agency in the United States responsible for anticipating and responding to both natural and manmade disasters, including acts of terrorism, hurricanes, tornadoes, and other disasters. FEMA has been criticized in the past for responding inadequately to disasters.

federal firearms license (FFL)   a federal firearms license in the United States granting authorization to manufacture or sell firearms.

Federal Judicial Center   the research and education center of the federal judicial system in the United States. Established in 1967, the Federal Judicial Center is controlled by a board chaired by the Chief Justice of the United States. It offers education and training programs for judges, attorneys, and non-judicial court employees, including personnel of clerk's and probation offices.

Federal Law Enforcement Training Center (FLETC)   a training facility located in Glynco, Georgia which offers law enforcement training for multiple federal agencies. Nicknamed Club Fed.

federal prison   a correctional facility operated by the federal government for the confinement of offenders found guilty of committing federal crimes. Federal prisons include those at Atlanta, Georgia and Marion, Illinois. See *Alcatraz*.

federal public defender   a public defender employed by the federal government. In addition to representing defendants in federal court, federal public defenders also work to obtain post-conviction remedies for convicted offenders, some of whom are facing the *death penalty*.

federal sentencing guidelines   a system that specifies penalties for various federal crimes.

Federal Witness Protection Program   see *Witness Security Program.*

feeblemindedness   defective mental abilities responsible for criminal behavior, according to English physician and criminologist *Charles Goring*. See *constitutional theory*.

fee splitting   the illegal practice by physicians of making unnecessary referrals to specialists, splitting the fee in return for the referral. See *white-collar crime*.

felicific calculus   same as *hedonistic calculus*.

felonious assault   an *assault* resulting in serious bodily harm of the victim. Felonious assaults often involve the use of a weapon. Compare with *aggravated assault*.

felony   a serious criminal offense that carries a term of one year of imprisonment or longer. Felonies include the most serious crimes, such as murder, rape, robbery, aggravated assault, grand theft, auto theft, and arson. Examples of less serious felonies are corruption of a minor, gross sexual imposition, forgery, embezzlement, passing bad checks, and manslaughter. Because they are more serious crimes, felonies generally result not only in more severe penalties, but also the loss of certain civil rights, including the right to vote and hold public office. Compare with *misdemeanor*.

felony disenfranchisement   the loss of rights, such as voting, as a result of a *felony* conviction. Felony disenfranchisement substantially limits the ability of felons to participate in a democracy.

felony murder doctrine   the doctrine which states that any death resulting from the commission of a felony constitutes murder, even if such death was incidental to the crime or accidental. Those involved in such a felony are thus charged with murder, despite the fact they did nothing purposeful or direct to bring about the decedent's death.

FEMA   abbreviation for *Federal Emergency Management Agency*.

female crime   crime committed by women and girls.

femicide   the killing of women and girls.

feminist criminology   criminology that challenges traditional, sexist conceptions of female offending and is also concerned with the victimization of women. Also, a school of thought in criminology that argues that women have been marginalized in the field of criminology, resulting in a largely male-dominated patriarchal discipline.

fence   a person who engages in *fencing*.

fencing   the illegal buying of stolen merchandise for the purpose of reselling it. It is argued that more than merely serving as an outlet for stolen goods, fences actually prompt thieves to steal. See *receiving stolen property*.

Ferracuti, Franco (1927–1992)   an Italian jurist and criminologist whose work bridged academic criminology and applied criminal justice. Ferracuti collaborated with other criminologists around the world, including *Simon Dinitz* and *Marvin E. Wolfgang*, with whom he coauthored *The Subculture of Violence*.

Ferri, Enrico (1856–1929)   an Italian criminologist associated with early positivist criminology. Ferri identified four types of criminals: insane, born, occasional, and those motivated by passion. He also suggested that criminal behavior was the result of both individual and environmental factors. See *Cesare Lombroso, positive school of criminology.*

fetal alcohol syndrome   physical and mental abnormalities in a fetus caused by the mother's heavy use of alcohol during pregnancy. Fetal alcohol syndrome is thought to contribute to later criminal involvement.

fetish   a sexual preference centered on a body part, inanimate object, or practice. Common objects of fetishes include certain types of clothing, fur, feet, and high-heeled shoes.

fetishism   the practice of engaging in a *fetish.*

feuding   an ongoing often violent conflict between two factions. One of the most famous examples of feuding was between the Hatfield and the McCoy families, which lasted for many years and resulted in numerous deaths.

FFL   abbreviation for *Federal Firearms License.*

fiber   any strand-like substance of interest in the investigation of crimes. Fibers of forensic interest include those from clothing, carpets, rugs, and wigs. See *hair and fibers.*

fiber optic spectrometer   a spectrometer that can be used by forensic scientists to analyze material.

Fifth Amendment   the amendment of the U.S. Constitution that guarantees fair treatment in the criminal justice system, including the right against *self-incrimination.*

filicide   the murder of one's child. See *Smith, Susan.* Compare with *fratricide, matricide, parricide, patricide, sororicide.*

financial fraud   intentional deception involving financial matters for personal gain. See *fraud.*

financial penalty   a penalty that requires a defendant to pay money. Financial penalties include *fines, day fines,* and *restitution.*

fine   an amount of money levied on a convicted offender in lieu of, or in addition to, another penalty. The range of fines is generally established by statute, giving the court *discretion* in the amount actually levied.

fingernail scrapings   the result of scraping the underside of human fingernails in order to collect possible *trace evidence,* such as blood or skin cells. See *bagging of hands.*

fingerprint   the residual pattern left on surfaces by the application of human hands and fingers. Fingerprints are unique to the individual, permitting their use in connecting suspects to crimes. See *arch, whorl.*

F

fingerprint card   a paper card on which inked fingers are pressed in order to create fingerprints.

fingerprint codes   identifying data on a fingerprint card, such as name, date of birth, sex, race, and Social Security number.

fingerprint identification   the process of determining the identity of an individual using fingerprints.

fingerprint pattern   the images of a fingerprint comprising arches, loops, and whorls.

fingerprint technician   an individual who specializes in lifting and analyzing fingerprints.

firearms examination   the in-depth examination of firearms suspected of operation in crimes.

firearms trafficking   the illegal distribution and sale of firearms.

fire debris analysis   the analysis of materials in a fire to ascertain the presence of substances used for ignition. See *arson, arson accelerant.*

fire setter   one who engages in *fire setting.*

fire setting   the tendency and attraction to start fires. Fire setting is one of the alleged three early indicators of future homicidal behavior, the other two are *bed wetting* and *cruelty to animals.*

firing squad   a method of *execution* in which several marksmen shoot the condemned to death. Typically one member of the firing squad will have a blank round, permitting all members to doubt that they actually caused the death of the condemned. Currently the firing squad is a possible form of execution in the state of Utah, as well as in a number of foreign countries. Compare with *gas chamber, hanging, lethal injection.*

First 48   an American television program on the A&E Network that depicts law enforcement officers investigating actual homicides. It derives its name from the belief among investigators that the first 48 hours are crucial to solving a homicide.

First Amendment   an amendment of the U.S. Constitution that guarantees freedom of religion and speech.

first offender   one who is accused, charged, or convicted of a crime for the first time. First offenders generally receive more lenient treatment by authorities. Compare with *chronic offender, repeat offender.*

first responder   any member of safety forces, including law enforcement, firefighters, or emergency medical technicians who are among the first to arrive at the scene of a crime, disaster, or other public emergency.

fix   to use influence to avoid or minimize the legal consequences of criminal, traffic, or other citations, or to unlawfully influence the outcome of sporting events or games of chance.

flagellation   same as *flogging.*

flailing   the practice of using a whip or other flexible weapon to beat convicted offenders. Compare with *caning.*

flash bang   a device that explodes with a loud noise and emits a blinding light to create a diversion in support of a *high-risk entry.*

flasher   one who exposes the genitals to others, usually for sexual gratification. See *indecent exposure.*

flat sentence   a sentence to confinement without a minimum or maximum period.

flaying   the practice of skinning. Some murderers have been known to flay their victims. See *Gein, Ed.*

fleeing   attempting to avoid capture by law enforcement authorities.

flick-knife   British for *switchblade.*

Fleiss, Heidi (1965–present)   a woman who ran a high-priced *call girl* operation in Los Angeles and became known as the Hollywood Madam. Fleiss was convicted of income *tax evasion.* See *prostitution.*

floater   a deceased person floating in water. Gases from *putrefaction* fill the body cavities, causing the body to rise to the surface and float.

flogging   a form of corporal punishment where the convicted offender is subjected to multiple lashes from a whip or flail. Compare with *caning, flailing.*

floodgate theory   the notion that the decriminalization of a certain behavior will dramatically increase its adoption. The floodgate theory applies to behaviors such as homosexuality and marijuana use.

fMRI   abbreviation for *functional Magnetic Resonance Imaging.*

focal concerns theory   a perspective advanced by criminologist Walter B. Miller which asserts that behavior of socially disadvantaged youths is focused around six concerns: trouble, toughness, smartness, excitement, fate, and autonomy.

folk crime   offenses resulting from social change and sentiments rather than because of other typical motives for crime. An example of this type of crime is the automobile assembler who leaves a beer can inside the door of the car he is helping to build in order to show dissatisfaction with some aspect of society.

F

folk devils   a person or group portrayed as deviant or undesirable and blamed for crime and disorder.

folkways   traditional habits of behavior, grounded primarily in meeting basic needs, as identified by early sociologist William Graham Sumner. Compare with *mores, norms.*

Food and Drug Administration (FDA)   the federal agency whose responsibilities include oversight of the quality and distribution of drugs in the United States.

footwear impression   an impression made in soil or other soft material by human shoes, boots, or other footwear. See *moulage.*

forced displacement   situation in which people are forced to flee their homes due to conflict, persecution, or violence. Also called forcible displacement and involuntary displacement.

forced entry   same as *break-in.*

forcible rape   another term for *rape.* Forcible rape, according to the *Federal Bureau of Investigation (FBI),* is one of the seven *index crimes.* Compare with *statutory rape.* See *sexual assault.*

Ford Foundation   a large private foundation that supports research on issues related to crime and justice.

Ford Pinto Case   an infamous case in which the Ford Motor Company sold its Pinto model despite evidence from crash tests that showed that rear end collisions would result in fuel tank explosions. It is estimated that 500 people died as a result of this type of collision. It was determined that Ford proceeded with the manufacture of the car after cost-benefit analysis suggested it would be cheaper to pay for the resulting lawsuits than to make the necessary modifications to the Pinto. See *corporate crime, white-collar crime.*

Foreign Corrupt Practices Act   a 1977 federal law that prohibits the payment of bribes to obtain business. The Federal Corrupt Practices Act was passed in the wake of a number of prominent bribery cases involving American corporations and foreign governments.

forensic accounting   the application of accounting methods to *forensic science.* Applications of forensic accounting include investigating financial crimes. See *creative accounting.*

forensic anthropologist   an anthropologist who specializes in the application of the science of physical or biological anthropology to the legal process. Forensic anthropologists participate in the identification of human remains.

forensic anthropology   the application of anthropological methods to forensic science.

forensic anthropometry   the measurement of the human body and its constituent parts for forensic purposes.

forensic archaeology   the application of archeological methods to forensic science.

forensic art   art as applied to the investigation of crime and the identification of unidentified subjects, including composite drawings, facial reconstruction, and the use of computer-assisted composite programs. Forensic art was used in the capture of *John List*.

forensic artist   an artist who specializes in drawings, paintings, or sculptures to help solve crimes. The work of forensic artist Frank Bender assisted in the capture of *John List*.

forensic autopsy   an *autopsy* conducted in a criminal case.

forensic botany   botany as applied to the investigation of crime.

forensic dentistry   the branch of dentistry focusing on legal issues, criminal issues in particular. See *forensic odontology*.

forensic engineering   engineering as applied to the investigation of crime. An application of forensic engineering is the analysis of the structure of a building believed to have collapsed due to shoddy construction.

forensic entomology   entomology as applied to the investigation of crime. Forensic entomologists can study maggot activity on human remains and estimate the approximate date of death. See *body farm*.

forensic error   mistakes made in the collection or processing of forensic evidence.

forensic facial reconstruction   the forensic specialty that involves recreating a human face from a completely or partially denuded skull for the purpose of identification. See *forensic art*.

Forensic Files   an American television program that depicts how forensic science is used to solve actual crimes.

forensic geology   the application of geological methods to forensic science. Forensic geologists study soil and rocks to determine their region of origin.

forensic interview   a sensitive, developmentally appropriate interview of a child to determine whether he or she has been abused or has witnessed a crime.

forensic linguistics   the application of linguistic methods to forensic science.

F

forensic mapping the use of mapping techniques to create and document a detailed visual record of crime scenes and other locations of forensic interest.

forensic medicine the branch of medicine dealing with legal issues, criminal issues in particular.

forensic nurse a nurse specially trained to collect evidence of crimes from victims. Compare with *Sexual Assault Nurse Examiner (SANE)*.

forensic odontology the analysis of dental structures in solving crime. Forensic odontologists can compare the teeth of homicide victims to x-rays on file. Sometimes, they can also extract genetic material for *DNA testing*. The scientific basis of forensic odontology has been challenged.

forensic pathologist a medically trained professional who is expert at determining the *cause of death*. Compare with *coroner, medical examiner*.

forensic psychiatrist a physician who specializes in *forensic psychiatry*.

forensic psychiatry the branch of psychiatry dealing with legal issues, criminal issues in particular.

forensic psychologist a specialist in *forensic psychology*.

forensic psychology the branch of psychology that deals with legal issues, including crime and justice.

forensic science the field concerned with the scientific detection and investigation of crime. Forensic science includes, but is not limited to ballistics, criminalistics, *hair and fibers*, and DNA testing.

Forensic Science Society see *The Chartered Society of Forensic Sciences*.

Forensic Specialties Accreditation Board (FSAB) a program that oversees other forensic specialty boards that certify forensic scientists.

forensic video analysis the analysis of video images in legal matters.

forest crime crime involving the illegal exploitation of forests, such as timber smuggling and harvesting endangered wood species. Compare with *wildlife crime*.

forfeiture see *asset forfeiture*.

forgery to create or alter money or documents for the purpose of deception or financial gain. See *naive check forger, systematic check forger*.

fornication sexual intercourse between two people who are not married to each other.

Fort Hood shooting the *mass murder* of 13 people and injury of many others at Fort Hood in Texas in 2009. Nidal Hasan, an Army major and psychiatrist, was responsible for the shootings.

Fourier transform infrared spectroscopy (FTIR)   a technique that yields an infrared spectrum of solid, liquid, or gas samples. FTIR is used to identify explosive materials and other samples of forensic interest.

Fourth Amendment   the amendment of the U.S. Constitution that prohibits unreasonable searches and seizures.

frame-up   an attempt to make it appear another party is responsible for a crime.

franchise fraud   a scheme in which the defrauded persons are led to believe they can make huge sums of money by purchasing what are really bogus franchises.

Fraternal Order of Police (FOP)   a large union of sworn law enforcement officers. The Fraternal Order of Police, with more than 325,000 members, promotes officer benefits and the protection of law enforcement officers.

fratricide   the killing of one's brother. Compare with *sororicide*.

fraud   the illegal use of deceptions or misrepresentations to gain an unfair advantage.

fraudster   one who engages in *fraud*.

fraudulent insurance claim   an attempt to obtain a benefit from an insurance company for a loss that did not legitimately occur. An example is pushing one's motor vehicle into a lake and reporting to the insurance company that it was stolen. See *fraud*.

freebase   a purified solid form of cocaine (such as crack) that is obtained by treating powdered cocaine hydrochloride with an alkaloid base (such as sodium bicarbonate) and that can be smoked or heated to produce vapors for inhalation.

Freedom of Information Act (FOIA)   a U.S. law which requires federal agencies to disclose non-classified or declassified documents to those entities who request them. It was through the FOIA that the public learned about former *Federal Bureau of Investigation* director *J. Edgar Hoover's* extensive files on celebrities and politicians.

free will   the notion that individuals in society, including criminals, freely choose their course of action. The notion of free will is found in the writings of 18th-century jurist *Jeremy Bentham* who asserted that humankind has the ability to choose a course or action, the choice generally motivated by hedonism. See *classical school of criminology, hedonistic calculus*.

Freudian theory   psychoanalytic theory advanced by Sigmund Freud. Freudian theory focuses on conflicts in the individual arising during childhood, some of which are thought to lead to criminal behavior.

friction ridge   the raised features on human hands and fingers that make possible fingerprints. See *fingerprint pattern*.

frisk   to manually search an arrestee or suspect for weapons, drugs, or other contraband. Frisking is necessary to protect the arresting officer. Officers have to be extremely careful not to get pricked by hypodermic needles, especially since they may carry infectious diseases. The practice of frisking by law enforcement is covered by the *Fourth Amendment.*

frotteurism   the practice of rubbing or touching another without consent for sexual pleasure. An example of frotteurism is the person who moves about in a crowded place, such as a bar or subway car, and takes advantage of the close quarters to make physical contact with another.

frustration-aggression hypothesis   a theory that argues *aggression* results from blocking an individual's goals.

Frye rule   the legal requirement that scientific evidence introduced into court must have acceptance by the larger scientific community. *Bite mark identification* is an example of evidence which might not stand up to the Frye rule because of its lack of reliability. On the other hand, *DNA testing* serves as an example of a technique which meets the standard. See *expert testimony, junk science.*

FTIR   abbreviation for *Fourier Transform Infrared Spectroscopy.*

fugitive   an individual who is fleeing arrest or prosecution.

Fugitive Safe Surrender   a program managed by the U.S. Marshals Service where fugitives are able to turn themselves in to law enforcement personnel at a church or other neutral location. This program saves *offenders* from worrying about eventual arrest and protects law enforcement officers from high-risk arrest situations. The offenders may or may not receive more lenient treatment as a result of surrendering. See *fugitive.*

fuming   a technique for making *latent prints* visible and usable.

functional family therapy   a program designed for young people aged 11 to 18 that addresses risk and protective factors both within and outside the family. See *Blueprints for Healthy Youth Development.*

functionalism   the sociological notion that deviant behaviors like acts of crime perform a necessary and useful function in society by reinforcing the norms prohibiting such behaviors.

furlough   a leave granted an inmate from a correctional facility or halfway house to pursue employment, education, treatment, or other legitimate activities.

Gacy, John Wayne (1942–1994) a prolific *serial killer* operating in and around Chicago, Illinois in the 1970s. Gacy, owner of his own construction firm, lured and murdered 33 young men, burying their remains under and around his house. He was convicted of 33 murders and executed in Illinois by the electric chair in 1994. While awaiting execution, Gacy, who had once played a clown earlier in his life, painted a number of self-portraits as a clown. See *serial murder*.

gag a piece of fabric or other material placed in or around the mouth to keep the individual from speaking or crying out.

gag order a judicial order or formal directive which forbids people from discussing something publicly.

Gainesville Ripper nickname given to the unknown *serial killer* that murdered five college students in Gainesville, Florida in the summer of 1990. *Danny Rolling* was eventually charged and convicted in the murders. See *serial murder*.

Gall, Franz Joseph (1758–1828) a German physiologist who developed cranioscopy that later evolved into *phrenology*. Gall proposed that the shape of the human skull gave clues to the mental and moral attributes of the individual. See *constitutional theory*.

Gallo, Joseph "Crazy Joe" (1929–1972) a member of *organized crime* in New York. Gallo was gunned down in a New York restaurant.

gallows the structure, most often constructed of two upright posts and a crossbeam from which is suspended the rope used to execute a condemned person by *hanging*. See *Tyburn tree*.

Galton, Francis (1822–1911) a Renaissance man whose contributions included advances in fingerprinting and examination of the inheritability of traits.

Gambino, Carlo (1902–1976) the *boss* of New York Italian *organized crime* for many years. Gambino was succeeded by *John Gotti*.

gambling engaging in games of chance.

gang   a group of individuals, often organized along racial or ethnic lines, whose members share values and a sense of identity. Gangs are associated with violence and other criminal activity. See *Bloods, Crips, Gangster Disciples, Latin Kings, security threat group*.

Gang Resistance Education And Training (G.R.E.A.T.)   a program developed by the *Bureau of Alcohol, Tobacco, and Firearms (BATF)* designed to help youths refrain from becoming involved in gangs. G.R.E.A.T. works in schools with law enforcement officers who teach a specific curriculum. See *gang*. Compare with *Drug Abuse Resistance Education (DARE)*.

gangster   a member of a criminal gang, particular one associated with *organized crime*.

Gangster Disciples   a Chicago-based *gang* reputed to be involved in a variety of criminal activity, including drugs and murder.

gaol   a term for a *jail* or *prison*.

Garofalo, Raffaele (1852–1934)   an Italian professor of criminal law who argued that criminality has an organic basis and therefore is inherited. See *positive school of criminology*.

garrote   a device consisting of a rope or wire with handles used to strangle a person.

gas chamber   a means of legal *execution* in which the accused is placed in an air-tight chamber after which sulfuric acid and cyanide are combined to form lethal gases which bring about death. See *death penalty*. Compare with *electric chair, firing squad, hanging*.

gas chromatography-mass spectrometry   a method of identifying specific chemical substances from samples. Applications of gas chromatography-mass spectrometry include the analysis and investigation of drugs, explosives, and arson.

gate fever   the emotional feeling attributed to prison inmates scheduled for release. Gate fever includes anxiety about where they will live, what they will do to earn a living, and whether they are able to refrain from engaging in crime.

gauge   the caliber of a *shotgun*, measured by the number of lead balls the size of the inside diameter of the barrel that comprise one pound. For example, a 12-gauge shotgun is called that because it requires 12 lead balls the size of the barrel's diameter to equal one pound. Compare with *caliber*.

gay bashing   the practice of assaulting gays or lesbians. A notorious case of gay bashing was the senseless murder of *Matthew Shepard* in Laramie, Wyoming in 1998. See *bias crime, hate crime*.

Gein, Ed (1906–1984) a *serial killer* and body snatcher, and fetishist who admitted to the murders of two women, but whose atrocities included fashioning household keepsakes from his victim's bone and skin. His crimes are believed to have inspired the Alfred Hitchcock movie *Psycho*.

gel electrophoresis the use of electricity to separate DNA fragments. Gel electrophoresis is an improvement over earlier gravity-based techniques.

gendarme one of a group of soldiers functioning as law enforcement officers in France.

gender specific services services for at-risk and delinquent girls that meet needs given their age and stage of development. Gender specific services address such issues as teenage pregnancy, eating disorders, body self-image, and sexually transmitted diseases.

general deterrence the characteristic of laws or policies that keep those contemplating crime from acting. General deterrence is essentially a legal threat. Compare with *specific deterrence*.

general rifling characteristics the patterns that appear on bullets as a result of passing through the rifled barrel of a firearm. See *General Rifling Characteristics File, grooves, lands, rifling*.

General Rifling Characteristics File a comprehensive file of *general rifling characteristics* maintained by the *Federal Bureau of Investigation (FBI)*.

General Social Survey (GSS) a survey of a representative sample of Americans conducted annually by National Opinion Research Center at the University of Chicago. The GSS occasionally includes items related to crime and justice. These data are archived at the *Interuniversity Consortium for Political and Social Research*.

General Strain Theory a revised version of *strain theory* advanced by criminologist Robert Agnew in which he specifies some of the strains that cause individuals to commit crime.

General Theory of Crime a criminological theory advanced by criminologists Michael Gottfredson and Travis Hirschi that posits that offenders are those who are low in self-control and who selfishly pursue their own self-interests. According to this theory, the factor most responsible for this is deficient parenting.

genetic fingerprint the unique genetic characteristics an individual possesses.

genetic transmission theory the theory that criminal tendencies are passed from one generation to the next through inherited traits. See *biocriminology*.

G

genocide the systematic killing of a people. An example of attempted genocide is the extermination of the Jews by the Nazis. See *crimes against
humanity, ethnic cleansing, Holocaust.*

Genovese, Kitty (1935–1964) a young New York woman who was repeatedly assaulted and eventually murdered in the presence of 38 witnesses,
none of whom came to her defense. Her murder gave rise to a series of
social psychological studies designed to understand this reluctance of
bystanders to intervene. See *bystander.*

geographic profiling the use of detailed geographic information and specialized analytical tools, such as the Rigel geographic profiling system, to locate
offenders. Geographic profiling can be used in any criminal case where a
suspect's movements can be plotted as points on a map. See *profiling.*

getaway car a motor vehicle used by offenders to evade capture after committing a crime.

get tough policies criminal justice policies intended to treat offenders more
harshly. Get tough policies include those that treat juvenile offenders as
adults and those that result in longer prison sentences with diminished
chance of parole. See *three strikes and you're out.*

gibbet in England, an upright post with an extended arm from which executed criminals were displayed to the public. The corpses were sometimes
left until they rotted, purportedly to serve as a deterrent to would-be
criminals.

Gillis, Lester (1908–1934) real name of Baby Face Nelson, a violent bank
robber and murderer of the 1930s. Gillis was eventually shot and killed by
law enforcement officers.

Gilmore, Gary (1940–1977) a convicted murderer who was the first person
to be executed after the U.S. Supreme Court ruled the *death penalty* to be
unconstitutional in the way it was administered in 1972. Gilmore's execution was also noteworthy in that he was put to death by *firing squad.* See
capital punishment.

Giteau, Charles (1841–1882) the individual who assassinated President
James A. Garfield. See *assassination.*

global crime crime of cartels and *terrorist* organizations that transcend
national borders to span the world.

global criminology the specialty within *criminology* that focuses on *global
crime.*

global drug trade the intercontinental sale and distribution of controlled
substances.

global fallacy   the tendency to explain all crime with a specific theory. The global fallacy is common in criminology. An example is to argue that a theory adequately explaining homicide also explains sex offending. It may be unrealistic to think that a single theory could explain all crime, even though there are criminologists who make this claim.

Glueck, Eleanor (1898–1972)   an American criminologist who with her husband Sheldon was a pioneer in conducting studies of delinquency. See *Glueck, Sheldon.*

Glueck, Sheldon (1896–1980)   a Polish-American criminologist who with his wife Eleanor was a pioneer in conducting longitudinal studies of delinquency. Glueck spent his career at Harvard University where he earned his Ph.D. His books include *500 Criminal Careers* and *Unraveling Juvenile Delinquency.* See *Glueck, Eleanor.*

gold investment fraud   a scam that is often perpetrated during tough economic times, which lures unsuspecting investors into buying gold, silver, or other precious metals.

good boys, bad boys   terms used to refer to the youths studied by *Walter C. Reckless* and *Simon Dinitz* in their effort to test Reckless's *containment theory.* School teachers in the study were asked to categorize boys in their classes as either good boys or bad boys.

good cop, bad cop   a tactic employed by law enforcement interrogators to elicit a *confession* from a criminal suspect. One officer plays hard-nosed investigator who is out to get the suspect while the other pretends to be the suspect's friend. The ploy sometimes results in a confession by the suspect. See *interrogation.*

good faith exception   an exception to the *exclusionary rule* that is permitted because law enforcement were operating in good faith.

good time credit   days taken off the sentences of convicted offenders for good behavior. In practice, prison inmates receive good time for the absence of infractions. Also good time credit. Compare with *bad time.*

Goring, Charles (1870–1919)   a researcher who compared a number of English convicts to college students, hospital patients, and soldiers in an effort to identify characteristics related to criminality. Though Goring did not find significant differences in the physiques of the two groups, he asserted that the convicts were mental defectives. Goring used subjective impressions to arrive at his conclusions about his subjects' mental ability that lacked the validity and reliability of standardized tests.

goths   youths who wear black clothing, listen to dark music, and otherwise separate themselves from conventional school groups and activities.

Suspicion was cast on goths in the wake of the *Columbine massacre* because Harris and Klebold were associated with that subculture.

Gotti, John (1940–2002)   an infamous *organized crime* figure of the late 20th century. Working his way up through the ranks of the *mob*, Gotti became known as the Dapper Don, because of his fashionable clothing, and later as the Teflon Don because criminal charges never stuck to him. He was finally convicted and sentenced to *life without parole*.

Gradations of Evil Scale   a scale consisting of 22 levels of criminal behavior, each of which conveys a unique level of violence and threat to the community. The Gradations of Evil Scale was developed by Columbia University forensic psychiatrist Michael Stone.

graduated sanctions   a series of alternatives for convicted offenders, each of which is slightly more punitive and restrictive than the last. Graduated sanctions, which give officials more options than simply sending an individual to jail or prison, can include such alternatives as *intensive supervision probation, house arrest,* and *electronic monitoring*.

graffiti   drawing, writing, or painting on buildings or other public surfaces by vandals or gang members. Graffiti is considered *vandalism* and it often signals the presence of gangs that use it to mark their territory. See *tagging, gang*.

graft   a form of political corruption or the unscrupulous use of a politician's authority for personal gain through bribery or similar practices in business or politics.

grand jury   a group of citizens, selected from the list of registered voters, who are brought together to consider evidence against criminal suspects charged with felonies. Grand juries operate in secrecy and issue an indictment when the evidence warrants the further pursuit of criminal charges. Compare with *petit jury*.

grand larceny   *theft* of goods or services exceeding a legislatively determined threshold that constitutes a *felony*.

grand theft   see *grand larceny*.

grand theft auto   grand larceny where the stolen article is a motor vehicle.

grass-eater   a corrupt police officer who accepts the occasional bribe. Compare with *meat-eater*. See *Knapp Commission*.

Great Train Robbery   the 1963 *robbery* of a mail train between Glasgow, Scotland and London, England. The Great Train Robbery netted the 15 robbers approximately £2.6 million.

green criminology   the branch of *criminology* that focuses on crimes against the environment and society's response to those crimes.

G

Green River killer the serial murderer responsible for the deaths of 48 women found in and around the Green River in Washington State during 1982 to 1984. *Gary Ridgway*, who was eventually convicted of the murders, confessed that he was responsible for additional victims.

grievous bodily harm serious physical injury as the result of a criminal act.

grifter one who conducts a *confidence game*.

grooming the process by which a sex *offender* draws a victim into a sexual relationship and maintains that relationship in secrecy. The shrouding of the relationship is an essential feature of grooming.

grooves in the *rifling* of the barrel of a firearm, the spiral troughs between the lands. See *lands*.

groper slang term for someone who inappropriately touches others in a sexual way. See *frotteurism*.

Gross, Hans (1847–1915) an Austrian jurist and magistrate credited with making significant early contributions to what is now known as *criminalistics*.

ground-penetrating radar a radar unit mounted on wheels that permits the user to analyze the substrata. Ground-penetrating radar is sometimes used to analyze disturbances in the earth to locate interred bodies.

group conferencing see *family group conference*.

group therapy psychodynamic therapy where clients participate in a group.

GSR abbreviation for *gunshot residue*.

Guantanamo Bay U.S. military base in Cuba where many terrorists have been confined.

Guardian Angels shortened name for The Alliance of Guardian Angels, Inc., a nonprofit organization of volunteers whose mission is to fight crime and provide positive role models for youths. The Guardian Angels was founded in 1979 by Curtis Sliwa in New York City. The distinctive red berets and tee shirts are meant to promote deterrence in areas known to have high rates of crime. Members are unarmed, but are trained in self-defense and may make a *citizen's arrest* if they see a crime being committed.

Guerry, Andre-Michel (1802–1866) a French lawyer and statistician who conducted some of the first empirical research in *criminology*. Guerry analyzed a variety of crime data, searching for regularities and patterns.

Guggenheim Foundation, Harry Frank a private foundation headquartered in New York City that awards grants to individuals for research on violence.

G

guillotine   a beheading machine consisting of a heavy slanted blade that drops between grooved uprights, used for executing condemned individuals, including convicted offenders. The guillotine, which first made its appearance in the late 1700s, was used as the form of *capital punishment* in France until the 1970s. During World War II, the Nazis used guillotines to execute thousands of prisoners. The guillotine is no longer in use. See *execution*.

guilt   the state of being legally responsible for committing a crime.

guilty   legally responsible for committing a crime. A person can be found guilty of a crime, but still be innocent.

guilty but mentally ill   a legal status where the accused is considered legally responsible for the criminal act for which he has been charged, but which acknowledges that his state of mind is compromised by mental problems. See *M'Naughten rule*.

gulag   an unpleasant prison for the confinement of political prisoners, generally associated with the former Soviet Union.

gun buy-back program   programs offering an incentive, usually cash, to turn firearms over to authorities. Such programs are based on the premise that fewer available firearms will result in fewer firearms related deaths and injuries. When such programs are offered, it is not unusual for them to yield so many firearms that officials run out of funds. Research on gun buy-back programs shows that they yield guns not typically used in the commission of crimes. While politically popular, they probably have little or no impact on the violent crime they are intended to prevent. See *gun control*.

gun control   the movement to restrict the purchase, possession, manufacture, distribution, or ownership of firearms by private individuals. See *Handgun Control, Inc., National Rifle Association*.

gun court   a *specialty court* designed to process mostly youthful offenders whose crimes involve the possession or use of firearms.

gun lobby   in general, organizations in the United States, such as the *National Rifle Association,* that work to preserve the *Second Amendment* of the Constitution of the United States. The gun lobby contributes large sums of money to politicians to ensure that the *right to bear arms* is preserved. See *gun control*.

gun meltdown   the practice of taking confiscated firearms and melting them in a foundry to prohibit their future circulation and use.

gun permit   written authorization issued by state or local authorities to permit the carrying of a firearm.

gun running   illegal *trafficking* in firearms. See *Operation Fast and Furious*.

gunshot   the audible report given off by a firearm. Also, the result of a firearm discharging.

gunshot detection sensor   an electronic device that makes use of sound waves to triangulate the origin of a *gunshot*.

gunshot residue (GSR)   the residue of chemicals that adheres to the hands and clothing as a result of firing a gun. See *paraffin test*.

gunshot wound   trauma to human or animal tissue caused by the impact of a projectile discharged from a firearm.

gun show loophole   the ability of individuals to sidestep firearm laws by purchasing firearms at a gun show.

gun violence   violence perpetrated with firearms.

Guzik, Jake (1886–1956)   a Jewish *organized crime* figure who worked for *Al Capone* in Chicago.

G

habeas corpus literally to take the body. Habeas corpus expresses the legal principle that states that a person not legally detained or confined must be freed.

habitual criminal see *habitual offender.*

habitual offender an offender who continues to engage in criminal behavior. Some states enact laws known as habitual offender statutes that define and are designed to control such offenders. See *chronic offender*.

hacker one who uses computer skills to gain entry to the computer system of another. In some cases, hackers simply do this as an intellectual challenge. In other cases, the hacker may either steal or corrupt information, or attempt to electronically transfer funds, as in the case of a financial institution.

hacking the unauthorized infiltration of a computer or system, often but not exclusively for malicious purposes. See *hacker*.

hair trigger the trigger of a firearm, most often a *pistol* or *revolver,* that requires relatively little pressure to bring about the firearm's discharge.

hairs and fibers a subspecialty of forensic science that focuses on the analysis of human, animal and synthetic hairs and fibers, such as those found in wigs, clothing, and carpeting.

halfway house a community-based correctional option where offenders released from prison make the transition back to community life by living in a house with other offenders. Halfway houses operate on the premise that convicts need a period of supervised treatment after release from prison to facilitate their adjustment on the outside. See *community corrections*.

hallucinogen a drug, natural or synthetic, that causes hallucinations. Hallucinogens can occur naturally, such as the psilocybin mushroom, or synthetically as in *LSD*.

Hamas a Palestinian resistance organization whose extremist members engage in *terrorism*.

handcuffs mechanical device that fastens around the wrists and is designed to restrain the person's mobility. See *belly chain, manacles*.

H

handgun   a small, easily concealed firearm designed to be held and fired with one hand. Because of their ability to be easily concealed and their disproportionate use in homicides and other violent crimes, handguns have been at the center of much controversy. See *Handgun Control, Inc.*, *pistol*, *revolver, Saturday night special.* Compare with *rifle, shotgun.*

Handgun Control, Inc.   an organization founded in 1974 that promotes gun safety legislation. Handgun Control, Inc. gained prominence with the involvement of Sarah Brady, wife of former Reagan White House Press Secretary James Brady, who was seriously wounded during an attempt on the President's life in 1981 by *John Hinckley.*

handgun permit   an official document allowing an individual to carry a handgun. See *carrying a concealed weapon.*

Handschu guidelines   a set of rules intended to regulate the police investigation of political activity in New York City.

handwriting analysis   the meticulous inspection of handwriting samples for comparison in criminal and other questionable cases. See *questioned documents examination.*

handwriting identification   see *handwriting analysis.*

hanging   a form of execution in which the condemned stands on a scaffold with a noose of rope tightened around the neck. When the trapdoor below the prisoner's feet is sprung, the condemned falls, breaking his or her neck. Compare with *electric chair, firing squad, gas chamber, lethal injection.*

hanging judge   a judge who has the reputation of meting out severe sentences for convicted offenders.

Hanssen, Robert (1944–present)   a former *Federal Bureau of Investigation (FBI)* agent who for 22 years spied for the Soviet Union against the United States. He is serving multiple consecutive life sentences in federal prison.

hardened criminal   an *offender* considered beyond rehabilitation or redemption.

hard labor   a form of punishment where the convicted *offender* must perform vigorous physical work as part of the *sentence.*

harm   the negative consequences of an act. Harm can be to an individual, an organization, or society as a whole.

harmful error   an error in handling a criminal case that compromises the constitutional rights of the defendant.

harmless error   an error made by a court that does not change the outcome.

harm principle   the notion that criminal law should exist only to prevent people from doing harm to others. Compare with *offense principle.*

Harris, Eric (1981–1999) one of the two perpetrators of the *Columbine Massacre*. After committing his crimes, Harris took his own life. See *Klebold, Dylan*.

Harvey, Donald (1952–present) a *serial killer* who murdered numerous people while working as a hospital orderly. He is currently serving the first of 28 consecutive life sentences at the Allen Correctional Institution in Lima, Ohio. See *angel of death, serial murder*.

hate crime a criminal offense motivated by hatred of a specific race, ethnic minority, or sexual orientation. The *Federal Bureau of Investigation* systematically collects data on hate crimes. In many states, hate crimes are codified as offenses distinguished from the core offenses, such as assault, vandalism, or intimidation. *Gay bashing* is an example of a hate crime. See *bias crime, Shepard, Matthew*.

Hate Crime Statistics Act a federal law enacted in 1990 which requires the collection of data on crimes committed as a result of the victims' race, religion, ethnicity, disability, or sexual orientation. See *hate crime, Shepard, Matthew*.

hazing rituals performed by Greek sorority and fraternity systems and other fraternal organizations as rites of passage. Hazing has been known to result in injury and death.

Head Start a well-funded, large-scale federal primary prevention program designed to give socially disadvantaged preschool children a boost through educational, health, nutrition, and other forms of assistance.

healthcare fraud see *Medicaid fraud, Medicare fraud*.

hearsay *testimony* covering information that is not considered direct knowledge by the witness and is therefore characterized as unreliable.

Hearst, Patty (1954–present) an heiress of the Hearst family who as a young woman was kidnapped by the *Symbionese Liberation Army*. See *kidnapping, Stockholm syndrome*.

hedonism the pursuit of pleasure and the avoidance of pain. See *classical school of criminology, hedonistic calculus*.

hedonistic calculus the computations individuals are said to make about both the possible pleasure and pain they will derive by engaging in specific behavior. According to the classical school of criminology, such individuals will choose the course of action which maximizes pleasure and minimizes pain. Also referred to as *felicific calculus*, hedonic calculus. See *Bentham, Jeremy*.

Hell's Angels headquartered in California, an organized group of motorcycle enthusiasts reputedly involved in criminal activity, including protection rackets, drug trafficking, murder for hire, as well as various property crimes. See *gang, outlaw motorcycle gang.*

hemp   slang for *marijuana.*

heritability studies   research studies that explore the genetic bases for traits and behavior, including criminal involvement. See *biocriminology.*

heroin   a narcotic derived from opium that is usually injected intravenously, inducing a euphoric state in the user. Heroin is highly addictive. Compare with *methadone.*

hesitation marks   found on some suicide victims, these are shallow, slashing-type wounds that run parallel to a deeper, fatal wound caused by a knife or other sharp object. These marks are generally indicative of second thoughts or hesitation in the final moments before the fatal injury is inflicted. When present on homicide victims, such marks may represent a form of torturing the victim before murder or possibly an attempt by a perpetrator familiar with the concept of hesitation marks to make a homicide appear to be a suicide. See *staged crime scene.* Compare with *defense wounds.*

hidden crime   crime not reported to authorities and therefore not represented in official crime statistics. Compare with *reported crime.* Also referred to as the *dark figure of crime.*

hierarchical linear modeling (HLM)   a set of statistical techniques which permits the simultaneous analysis of variables representing different levels of analysis. For example, in the analysis of delinquency in a community, HLM is capable of examining the effects of individual, family, and community factors as well as assessing changes over time in longitudinal data.

hierarchy rule   the *Uniform Crime Reports (UCR)* rule that specifies that only the most serious crime in a complex offense is recorded. For example, if in a single incident a suspect commits murder, robbery, and rape, only the murder is recorded for UCR purposes since it is the most serious crime. This is no longer the case with the *National Incident-Based Reporting System (NIBRS),* which captures all offenses in a single incident.

high explosives   explosives with high detonation velocities. Examples of high explosives include dynamite, plastic explosives, and TNT. Compare with *low explosives.*

high-risk entry   an entry by law enforcement officers where there is a high probability of armed resistance by the occupants, which in turn poses a greater risk to the officers. See *flash bang, S.W.A.T.*

high sheriff   same as *sheriff.*

high-speed pursuit the practice by law enforcement officials of engaging in motor vehicle chases of suspects at high rates of speed to effect the suspect's capture. High-speed pursuits are controversial because they have resulted in the deaths and injuries of officers, suspects, and innocent persons. Specialized training in high-speed pursuit is designed to reduce such deaths and injuries. Using devices like the *Stop Stick* can reduce the need for high-speed pursuits.

high-velocity blood spatter *blood spatter* created by the impact of a weapon, such as a bullet or blunt object on human tissue. Compare with *castoff.* See *blood spatter analysis.*

highwayman a now dated term from the Elizabethan era until the early 19th century used to describe an individual who robbed and sometimes murdered travelers. See *robbery.*

highway patrol law enforcement agency whose primary role is the enforcement of motor vehicle laws in a state. Highway patrol officers typically have more limited powers than *state police.*

hijacking a crime in which a carrier, most often a motor vehicle or aircraft, is waylaid, diverted, or otherwise illegally controlled for criminal or terrorist purposes. See *air piracy, carjacking, piracy, terrorism.*

Hillside Strangler nickname given to the person or persons who kidnapped, raped, tortured, and killed a number of women and girls near Los Angeles in the late 1970s. *Kenneth Bianchi* and *Angelo Buono* were eventually deemed responsible for the crimes. See *serial murder.*

Hinckley, Jr., John (1955–present) a man who unsuccessfully attempted to assassinate President Ronald Reagan in 1980. Hinckley's motives for the attempted *assassination* were tied up in an obsession with actress Jody Foster and her role in the movie *Taxi Driver.* Hinckley remains confined in a mental institution.

histology the study of human tissue. Specialists in histology conduct examinations of cells.

Historical Violence Database a large, multifaceted database of violent incidents in the United States, including homicides, suicides, and serious assaults going back several hundred years. The Historical Violence Database is maintained at the Criminal Justice Research Center at The Ohio State University.

hit a murder for hire or by order. Hits are normally, but not exclusively associated with *organized crime.* See *contract, hit man.*

hit man a person who performs murder for hire or is under the orders of a superior as seen in organized crime. See *Murder, Inc.*

HITS   abbreviation for the *Homicide Investigation Tracking System*.

hoax   a purposeful deception motivated either by humorous or malicious intent.

Hobbs Act   federal anti-racketeering legislation in the United States intended to control interference with interstate commerce. See *racketeering*.

holding cell   a cell designed to temporarily hold an *arrestee*.

hole   an unpleasant, dungeon-like cell in a prison reserved for recalcitrant convicts. See *solitary confinement*.

Hollywood Madam   nickname for *Heidi Fleiss*.

Holocaust   the systematic attempt in the 1930s and 1940s by the Nazis to exterminate anyone of Jewish ancestry. The Holocaust resulted in the deaths of more than six million Jews from 1939 to 1945. See *concentration camp, death camp, genocide, war crimes*.

holy trinity   the group of three elements thought to be critical for a criminal conviction, consisting of *physical evidence*, witnesses, and a *confession*.

home confinement   same as *home detention*.

home defense   actions taken to defend one's place of residence.

home detention   the practice of ordering an offender to remain at his or her place of residence as an alternative to confinement in prison. Home detention can be supplemented with *electronic monitoring*.

home invasion   the forced entry into a home for the purpose of committing *robbery, rape,* or otherwise terrorizing the residents.

homeless persons   persons, most often unemployed and frequently suffering from mental illness, who do not have stable residences. Homeless persons are commonly perceived as threatening, even though they are responsible for little actual crime. See *public order crime*.

Home Office   the government agency in the United Kingdom responsible for criminal justice research and policy. The Home Office is headquartered in London, England.

Home Office Counting Rules for Recorded Crime   a set of standards for the collection and reporting of crime data for England and Wales. Compare with *Uniform Crime Reports*.

home security   any device or program intended to protect homeowners against crimes like burglary and theft.

homicide   the taking of human life by another. While often associated with criminal behavior, homicide also includes the lawful taking of life, such as

H

justifiable killings of criminal suspects by police, and instances of self-defense by citizens. See *murder, Supplementary Homicide Reports (SHR)*.

Homicide Investigation Tracking System (HITS)   a program operated by the Attorney General of the State of Washington to track and investigate homicides and other violent crimes with similar methods of operation and patterns. HITS, which contains data from more than 10,000 homicides and 8,000 sexual assaults, is a contributor to the *Violent Criminal Apprehension Program (VICAP)* database of the *Federal Bureau of Investigation (FBI)*. The HITS staff also consult with other jurisdictions on homicide and rape cases.

homicidomania   the aberrant desire to commit murder.

homosexuality   the practice of sexual relations between persons of the same sex. There are still places where homosexuality is a crime.

honor killing   the killing of a family member who is thought to have brought shame on the family.

hood disease   term used to describe the stressful experience of young people who have lived in a disadvantaged, violent neighborhood. Compare with *posttraumatic stress disorder*.

hoodlum   a street offender, especially a young one. Also shortened to hood.

hooker   a *prostitute*. See *call girl, Fleiss, Heidi*.

hooligan   one who engages in *hooliganism*.

hooliganism   disorderly conduct or disruptive and unlawful behavior, such as rioting, bullying, and vandalism. Also, the often unprovoked violence associated with soccer and other European athletic events. Hooliganism has resulted in numerous deaths and serious injuries.

Hoover, J(ohn) Edgar (1895–1972)   the first and longtime director of the *Federal Bureau of Investigation (FBI)*.

horizontal prosecution   prosecution where different prosecutors handle each new step throughout the court process. Compare with *vertical prosecution*.

hormonal imbalance   a condition where an excess or insufficient amount of a particular hormone is said to be responsible for criminal behavior.

hostage   a person who is held until *ransom* is paid or other *conditions of release* are met. See *kidnapping, Stockholm syndrome*.

hostage negotiation   the art of working to bring about a peaceful resolution to a *hostage* situation. Those trained in hostage negotiation respond when a person is barricaded with a hostage or is threatening to commit suicide.

H

hostile witness   a witness in a criminal case whose testimony contradicts the case of the side that subpoenaed them.

hot spot mapping   the manual or computer-assisted placement on a map of *hot spots*.

hot spot policing   a law enforcement strategy that focuses resources on *hot spots* because of their disproportionate contribution to overall crime.

hot spots   geographical areas where there are a disproportionately high number of reported crimes. Hot spots can be used to allocate law enforcement resources. See *crime mapping, geographic profiling.*

house arrest   see *home detention.*

Howard, John (1726–1790)   an English philanthropist and prison reformer of the 18th century. See *Howard League for Penal Reform.*

Howard League for Penal Reform   named for prison reformer *John Howard*, the Howard League for Penal Reform is an organization that works with parliament, the media, criminal justice professionals, and the public to reduce crime and bring about meaningful change in the criminal justice system, including prisons.

human ecology   the study of the relationship between people and the physical space they occupy.

human identification   the art of establishing the identity of deceased persons by analyzing and reconstructing their remains. See *forensic anthropology.*

human rights   rights considered so universal that they belong to all persons. See *human rights violations, Human Rights Watch.*

human rights violations   violations such as *genocide* and *torture* that offend the sensibilities of most civilized societies.

Human Rights Watch   an international organization dedicated to protecting human rights around the world. Human Rights Watch, headquartered in New York, conducts investigations into alleged human rights violations, periodically publishing statistics on executions, tortures, and other violations. Ending the sexual abuse of prisoners has been one of its focus issues. It is supported by contributions from other organizations and individuals. See *human rights violations.*

human sacrifice   the practice of ritualistically killing a person, often for religious reasons. See *satanic cult.*

human trafficking   the trade in humans primarily to exploit them for purposes of forced *prostitution.*

hunger strike   a protest by jail or prison inmates during which they eat little or no food until their grievances are heard or conditions are improved.

hung jury   a *jury* unable to reach a unanimous *verdict* in a criminal case. Cases ending with a hung jury may be retried.

husband beating   assaults on a man by his wife. See *domestic violence.* Compare with *wife beating.*

hypnosis   a sleep-like state where a person is subject to suggestions from a hypnotist. Law enforcement authorities sometimes use hypnosis to help witnesses to a crime recall important details of the crime.

hypothalamic region   the part of the human brain that controls body temperature, thirst and hunger, among other functions. The hypothalamic region plays a part in controlling *aggression.*

H

ICPSR abbreviation for the *Inter-University Consortium for Political and Social Research.*

identification parade same as *lineup.*

Identikit a kit consisting of a myriad of facial features on overlapping transparencies used by law enforcement to identify suspects, witnesses, and other unknown persons of interest in investigations, including criminal and missing persons.

identity crime theft and associated fraudulent misuse of personal information, such as date of birth, Social Security number, credit card numbers, passwords and so forth.

identity theft the assumption by an offender of another's identity through the use of the victim's personal data, such as Social Security number, credit card number or other identifying data.

ignition interlock a device connected to the ignition of a motor vehicle that prohibits an intoxicated driver from starting the car. Ignition interlock devices are frequently used for those offenders who have been convicted of *driving under the influence.*

I-level a classification system for the Interpersonal Maturity Level of delinquents. Based on a continuum of cognitive development, the I-level has been used for diagnosis and treatment of youths in institutions.

illegal alien one who enters and remains in the country without proper authorization. Also referred to as undocumented alien.

illegal immigrant same as *illegal alien.*

illegal wire transfer sending money via wire transfer in connection with a crime. Criminals use illegal wire transfers to launder money because they lack a solid *audit trail.* See *money laundering.*

imitator same as *copycat.*

Immigration and Naturalization Service (INS) former name of *United States Citizenship and Immigration Services (USCIS).*

immunity from prosecution protection against criminal prosecution extended to those willing to testify against accomplices.

impeachment the official censure and removal from office of an official found to have engaged in criminal behavior or other serious misconduct.

impersonating an officer a criminal offense where the *offender* illegally poses as a law enforcement officer.

importation model the model which suggests that the subculture within a prison stems from both what happens in prison as well as what inmates bring to the experience. Compare with *deprivation model*.

importuning soliciting a person under the age of 13 for sexual purposes.

impression evidence *evidence* created by an object pressing against another, thereby transferring the characteristics of one onto the other.

imprisonment the punishment that consists of confinement in a prison. Imprisonment is among the more severe sanctions meted out by a *judge* at *sentencing*.

improvised explosive device (IED) a homemade bomb made of military explosives, such as artillery shells rigged to detonate either manually or remotely. IEDs are frequently used by terrorists abroad.

imputation of data the insertion of fictitious values in a data set as substitutes for missing values in order to perform data analysis.

in camera held privately in a judge's private chambers rather than in open court.

incapacitation one of the four primary purposes of punishment concerned with physically restricting the offender's ability to reoffend. Compare with *deterrence, rehabilitation, retribution, selective incapacitation*.

incarceration confinement in a jail, prison, or other detention or correctional facility.

incest sexual relations between persons who are related.

incidence the number of times a crime has occurred. Compare with *prevalence*.

incised wound a wound caused by a knife or other cutting instrument. Compare with *chop wound*. See *sharp force trauma*.

income tax evasion the intentional failure to report taxable income at the time of filing federal or other tax returns. Income tax evasion in the United States is a felony. It was income tax evasion that resulted in the conviction of *Al Capone*.

incompetent   the state of being unable to stand trial in a criminal case due to inability to understand the court proceedings and assist in the defense.

incorrigible   beyond the normal control of parents and caregivers. Incorrigible is used as an official status for youths who do not respond to authority. See *status offense*. Compare with *unruly*.

incriminate   to imply, suggest, or produce evidence which tends to prove that one is guilty of a crime. See *self-incrimination*.

incriminating evidence   evidence which could help *incriminate*.

inculpatory evidence   evidence that reinforces the alleged guilt of the accused. Compare with *exculpatory evidence*.

indecent   sexually explicit or that which offends the sensibilities of the average person.

indecent exposure   a criminal offense, most often a misdemeanor where the offender exposes genitalia or other sexual organs to others. Indecent exposure is generally defined in the law as a *misdemeanor*.

indefinite sentence   same as *indeterminate sentence*.

independent variable   in statistical analysis, any variable or factor that helps explain the phenomenon of interest. For example, if analyzing the causes of crime, "associating with delinquent peers" might be an independent variable. Compare with *dependent variable*.

indeterminate sentence   a sentence consisting of a range of months or years, the actual time served depends on considerations, such as the offender's rehabilitation, subtraction of good time, the discretion of the parole board and other factors. For example, a sentence of 7 to 25 years. Also referred to as *indefinite sentence*. Compare with *determinate sentence*.

index crimes   according to the *Federal Bureau of Investigation (FBI)*, the seven most serious crimes, including murder and non-negligent homicide, rape, robbery, aggravated assault, burglary, larceny, and motor vehicle theft. Under certain circumstances, arson is included as an index crime. Also referred to as index offenses. See *modified index, Part I offenses, Part II offenses, Uniform Crime Reports (UCR)*.

Indian Society of Criminology   a professional organization for criminologists in India.

indict   to formally charge, generally with a *felony*, based on the findings of a *grand jury*.

indictment   a document issued by a grand jury charging a person with a crime. Also called true bill. Compare with *bill of information*.

indigenous crime a crime committed by people who are native to a particular geographical location.

indigenous justice the involvement of indigenous people in the criminal justice system.

Indigenous Justice Clearinghouse a clearinghouse maintained by the Australian government for the purpose of disseminating information about *indigenous justice* to policymakers and other interested parties.

indigenous offender an offender that is a member of an indigenous people. Examples are Aboriginal people in Australia and Native Americans in Canada and the United States.

indigenous overrepresentation the disproportionate percentages of indigenous people having contact with the criminal justice system. Compare with *minority overrepresentation*.

indigent with respect to accused offenders, those without sufficient funds to retain private counsel. See *legal aid society, pro se, public defender*.

indigent defense the defense of individuals without the means to hire private counsel. See *public defender*.

individual characteristics characteristics of *evidence* that are unique to the object. Compare with *class characteristics*.

individual deterrence same as *specific deterrence*.

industrial espionage the illegal act of gathering and using information from private corporations by their competitors or others. See *espionage*.

industrial spying see *industrial espionage*.

inebriate a drunk person, most often one in public. See *public intoxication*.

ineffective assistance of counsel a claim raised by a convicted criminal defendant that their attorney's performance was so ineffective that it deprived them of their rights.

infanticide the killing of an infant. Infanticide most often is committed by parents or caregivers. Compare with *neonaticide*. See *homicide*.

informant one who supplies information to authorities about alleged crimes, often in return for payment or consideration in their own pending legal problems. Compare with *confidential informant*.

in forma pauperis a designation that denotes a person is indigent and therefore excused from paying certain fees, such as court filing costs.

information see *bill of information*.

informed consent   agreement to voluntarily participate in a research study with full knowledge of its purposes and possible physical and psychological consequences. At one time researchers conducted studies of prison inmates and others without obtaining informed consent. This practice is no longer acceptable.

infrared imaging   the use of infrared energy in forensic investigations to detect and identify attributes of interest without destroying the *evidence*.

inhalant   a substance inhaled for its mind-altering effects.

initial appearance   in adult courts, the first appearance of an offender before a judge or magistrate. At initial appearance, the accused is apprised of the charges and a bond is set. Compare with *preliminary hearing*.

injunction   a judicial order prohibiting a behavior or compelling one.

inmate classification   the practice of assigning prison inmates based on such factors as perceived risk or treatment needs.

inmate suicide   the taking of one's own life by a resident of a jail, prison, or other detention or correctional facility. Inmate suicide has been studied extensively by the *National Center on Institutions and Alternatives (NCIA)*.

Innocence Project   a national project committed to assisting confined offenders whose wrongful convictions might be overturned with DNA *evidence*. The Innocence Project, which is headquartered in New York City, is also dedicated to preventing future miscarriages of justice. See *convicted innocent*, *DNA testing*.

innocent   without guilt. Compare with *not guilty*.

innocent bystander   a *bystander* who has nothing to do with a crime, but who gets injured or killed during the commission of a crime.

Innovative Neighborhood Oriented Policing (INOP)   a program designed by the *Bureau of Justice Assistance (BJA)* to further community-based *demand reduction* for drugs. Piloted in eight U.S. cities, INOP is grounded in the importance of community, including *community policing* practices.

inquisitorial justice   justice where the court plays an active role with the prosecution in investigating the facts of the case. Compare with *adversarial justice*.

insanity   the state of being of unsound mind. The term when used in criminal justice relates to the defendant's ability to distinguish right from wrong. See *M'Naughten rule*.

insanity defense   a strategy employed by a defense counsel where they maintain the defendant was mentally ill at the time of the alleged offense, and consequently did not know right from wrong.

insect activity   activity by insects, including flies and beetles, on human and animal remains. Specialists can use insect activity to estimate how long a body has been dead. See *body farm, forensic entomology.*

Inside-Out Program   a national program where college students and prison inmates are brought together under the assumption that the experience will be mutually beneficial.

insider trading   the illegal use of confidential stock market information to one's own advantage. Notable cases of insider trading include those involving *Ivan Boesky* and *Michael Milken.* See *white-collar crime.*

inspector general   an official whose responsibilities include investigating corruption and other forms of governmental misconduct.

instant offense   the current offense with which a *defendant* is charged.

Institute for Juvenile Research (IJR)   a treatment and research center in Chicago with a long history of contributions to the study of juvenile delinquency. Early criminologists Clifford Shaw and Henry McKay were affiliated with the IJR and from there helped launch the *Chicago Area Project.* The IJR is now affiliated with the Department of Psychiatry at the University of Illinois at Chicago.

Institute for Youth Development (IYD)   a nonprofit organization that promotes the avoidance of alcohol, drugs, sex, tobacco, and violence. Central to IYD's philosophy is that youths are capable of making appropriate choices, especially if supported by their families. The IYD is headquartered in Washington, D.C.

institutionalization   the process that occurs in people during prolonged periods of confinement, resulting in their inability to cope with mainstream life outside the institution, and sometimes results in their desire to remain confined. Compare with prisonization.

instrumental crime   a crime such as *robbery* or *theft* that yields a demonstrable benefit for the perpetrator. Compare with *expressive crime.*

instrumental Marxism   the Marxist notion that the crime problem stems more from the process of criminalization than from individual or environmental factors emphasized in traditional Western theories. See *Marxist criminology.*

insurance fraud   a scheme in which a bogus claim is made to an insurance company based on a contrived loss. Insurance companies estimate their losses to insurance fraud in the billions of dollars annually. See *fraud.*

intake   the point at which a youth formally becomes involved in the juvenile justice process. Generally, at intake a youth is either released to parents or guardians or confined in a *detention center.*

integrative shaming the practice of making offenders take responsibility for their actions without public humiliation or labeling. See *disintegrative shaming, labeling perspective, shaming penalties.*

integrative theory a criminological theory which seeks to combine and reconcile two or more existing theoretical perspectives into a more comprehensive unified perspective.

intelligence-led policing a stance of law enforcement characterized by the proactive collection and use of intelligence data to prevent and control crime.

Intelligence Project, The a project of The Brookings Institution that explores the relationship between intelligence gathering and both successes and failures of public policy.

intensive supervision probation (ISP) a form of probation that involves an extra measure of supervision and control by the probation officer. ISP often is used for chronic and other high-risk offenders who pose a greater probability of reoffending. Also referred to as intensive probation supervision (IPS).

intent the purposeful mindset to undertake a certain course of action. Intent is an important element in establishing whether a person knowingly and willingly committed a crime.

intentional injury a physical injury caused through purposeful action by another. *Assault* is an example of an intentional injury. The *Centers for Disease Control and Prevention* analyze intentional injuries to determine the causes and correlates of violence.

interaction effect the results of one variable acting in concert with another such that together they produce an effect different from what either would produce individually.

interdiction efforts to stem the sale or distribution of illegal goods, most often narcotics and other drugs. Interdiction includes such activities as intercepting motor vehicles, watercraft, and aircraft that are transporting illegal drugs. See *supply reduction.*

intermediate sanction a punishment for a crime between a mild sanction and a severe sanction. Examples of intermediate sanctions include *electronic monitoring* and *intensive supervision probation (ISP)* where the more mild and severe extremes might be *non-reporting probation* and confinement in *prison*, respectively.

internal affairs a unit within law enforcement agencies responsible for investigating alleged misconduct and corruption. Internal affairs units have the reputation, deserved or not, of aggressively pursuing "dirty cops," at times using questionable methods to achieve their ends.

internalizing behavior   negative behaviors that are inner directed, such as blaming one's self or withdrawing socially. Compare with *externalizing behavior*.

Internal Revenue Service (IRS)   part of the U.S. Department of the Treasury, the federal agency responsible for enforcing provisions of the Internal Revenue Code. See *revenuer, tax evasion*.

International Association for Investigation (IAI)   the world's oldest and largest forensic organization, the IAI represents a diverse membership and promotes education, research, and the sharing of knowledge in the forensic sciences. See *forensic science*.

International Association of Chiefs of Police (IACP)   an organization of police executives from around the world. The IACP's activities include advocacy, training, and research.

International Crime Victim Survey (ICVS)   a series of surveys designed to measure respondents' experience with crime, policing, crime prevention, and fear of crime. The ICVS was administered in both developed and developing nations. See *victimization survey*.

International Criminal Court   a treaty-based court that serves the international community by adjudicating serious crimes, such as *genocide, war crimes*, and *crimes against humanity*.

international terrorism   *terrorism* that transcends the boundaries of any one country.

Internet crime   crime facilitated by use of the Internet.

Internet predator   a *sexual predator* who uses the Internet to identify and lure potential victims.

INTERPOL   acronym for International Criminal Police Organization. A multinational collaborative of law enforcement agencies that cooperates in the investigation and apprehension of wanted criminals. INTERPOL is headquartered in Paris, France.

interracial crime   crime occurring among individuals of different racial groups. Compare with *intraracial crime*.

interrogation   the informal or formal questioning of suspects or witnesses. See *good cop, bad cop*.

interstate compact   a written agreement among states and territories. Interstate compacts permit states to transfer parolees from one jurisdiction to another.

Interuniversity Consortium of Political and Social Research (ICPSR)   a membership consortium of colleges and universities established in 1962 and headquartered at the University of Michigan at Ann Arbor. The ICPSR maintains an extensive archive of social science data sets. It also offers training in quantitative methods, as well as technical assistance to researchers using ICPSR data. The *National Archive of Criminal Justice Data* is maintained at ICPSR.

intimate handler   an individual who exercises social control over another through emotional attachment. See *routine activities approach.*

intimate partner homicide   *intimate partner violence* that results in death.

intimate partner violence   violence between those living together who may or may not be related. See *domestic violence, intimate partner homicide.*

intraracial crime   crimes in which the offender and victim are of the same race. Compare with *interracial crime.*

intruder   a person who purposefully intrudes or breaks and enters into premises, especially with criminal intent.

inventory shrinkage   the loss of a company's inventory due to *employee theft*. See *pilfering.*

investigation   a formal process where available information about an alleged crime is collected and studied.

investigative profiling   see *profiling.*

investigator   a person who conducts an i*nvestigation*. Compare with *detective.*

involuntary manslaughter   the unintentional killing of another through recklessness or while committing a non-felonious illegal act. An example of involuntary manslaughter is accidentally killing someone while driving drunk. Also negligent manslaughter. Compare with *voluntary manslaughter.*

involvement   one of four elements of criminologist Travis Hirschi's control theory of delinquency. See *attachment, belief, commitment.*

Iran-Contra case   a scandal during President Reagan's administration where the U.S. government sold arms to Iran in exchange for the safe return of hostages. The money received from Iran was then used to fund the Nicaraguan contras. Both the sale of the arms and the giving of money to the contras were against the law. Also referred to as Irangate.

irresistible impulse   phrase used to describe the alleged inability of an individual to control the desire to engage in criminal behavior due to a mental disease.

Irwin, John (1929–2010)   a former convicted felon who became a respected criminologist. Irwin was a vocal proponent of *convict criminology.*

ISIS   abbreviation for *Islamic State in Iraq and Syria.*

Islamic State in Iraq and Syria (ISIS)   an al Qaeda splinter group that strives to create a Sunni Islamic state in portions of Iraq and Syria. ISIS is known for its extreme brutality, including *mass murder* and beheadings.

Islamophobia   negative reaction toward Muslims as a result of Islamic terrorist activities.

isolation unit   a unit in a prison, jail, mental hospital, or other institution where inmates or patients are segregated from the general population. Compare with *hole.*

Ivins, Bruce (1946–2008)   an American microbiologist believed by many to be the perpetrator of the anthrax attacks. After learning that he was being charged in the crimes, Ivins took his own life with a *drug overdose.* See *Anthrax case.*

jack roller a person who robs drunken or sleeping persons. A young jack roller in early 20th century Chicago was the subject of the life history and book *The Jack Roller*, by criminologist *Edwin H. Sutherland*.

Jack the Ripper the name of the *serial killer* who was responsible for the mutilation murders of five prostitutes in London's White Chapel district in 1888. The murders were never officially solved, despite numerous theories on possible suspects. See *serial murder*.

jail a short-term facility for holding accused or sentenced offenders. Compare with *lockup, workhouse, prison*.

jail break an escape from a jail or similar holding facility.

jailhouse lawyer a jail or prison inmate who becomes familiar with the law and advises other inmates on legal matters.

jail suicide the suicide of an inmate or detainee occurring in a jail. See *inmate suicide*.

Jeffrey, C. Ray (?–2007) an American criminologist best known for his work on *crime prevention through environmental design* (CPTED). Jeffrey spent most of his career as a faculty member at Florida State University.

J.F.K. assassination the *assassination* of President John F. Kennedy by *Lee Harvey Oswald* on November 22, 1963, in Dallas, Texas.

jihad a war fought by Muslims to preserve or spread their religious beliefs.

john a man who solicits sexual services from a *prostitute*.

John Adams Project a project whose mission was to support the legal representation of detainees at Guantanamo Bay, Cuba. It was discontinued in 2009.

Joliet the site of the famous Stateville Correctional Center in Illinois. Infamous inmates who served time at Joliet include *John Wayne Gacy* and *Richard Speck*.

J

Jones, Jim (1931–1978)   the founder and religious leader of the Peoples Temple who orchestrated the suicide or murder of more than 900 men, women, and children in Jonestown, Guyana. See *mass murder*.

Jonestown Massacre   the murder and suicide of more than 900 men, women, and children at the Peoples Temple in Jonestown, Guyana in 1978. See *Jones, Jim, mass murder*.

joyriding   the act by juveniles of taking and driving a motor vehicle without permission. Joyriding most often differs from *auto theft* in that those doing it generally intend to return the vehicle after they have their fun.

Judd, Winnie Ruth (1905–1998)   an Arizona woman whose murder and dismemberment of two women over romantic attachments received a lot of media attention, some of it related to the *death penalty*. Judd was judged incompetent to stand trial and spent years in mental hospitals from which she escaped multiple times.

judge   a governmental officer appointed or elected and invested with the authority to decide questions of law. Compare with *justice, magistrate*.

Judicial Conference of the United States   a federal conference whose role is to frame policy guidelines for courts in the United States.

judicial corporal punishment   *corporal punishment* as a court-imposed sentence.

judicial discretion   the inherent ability of a judge to make subjective decisions about cases before the court.

judicial immunity   immunity that protects the judiciary from legal action arising as a result of their decisions and actions, regardless of the severity.

judicial misconduct   misconduct perpetrated by judges in the official performance of their judicial duties. One of the best known cases of judicial misconduct was that uncovered by the *Federal Bureau of Investigation* in their *Operation Greylord* in Cook County, Illinois.

judicial release   a release of a convicted offender from prison prior to their release date with the balance of the sentence served on probation. Judicial release was previously known as *shock probation*.

Jukes family   a family studied by the sociologist Richard Dugdale in the late 19th and early 20th centuries to examine a number of factors, including criminality. The study was used in attempts to illustrate the inheritability of criminal tendencies and to support *eugenic criminology*. See *Kallikaks, nature-nurture debate*.

JUMP   abbreviation for *Juvenile Mentoring Program*.

junk science   practices used in the criminal justice process that have questionable scientific validity.

jurisdiction   an area with geographical boundaries where law enforcement, prosecutors, courts or other criminal justice functionaries have authority.

jurisprudence   the study of the nature of law and legal systems.

Juristat   a sophisticated computer program that permits attorneys to predict the behavior of legal actors, such as judges and juries, based on detailed historical data. See *jury consultant.*

juror   a citizen selected, usually from the list of registered voters, to decide the guilt of a defendant in a criminal case based on evidence presented.

jury   a body of citizens, usually selected from the pool of registered voters, charged with making determinations of guilt in a criminal trial based on presented evidence and testimony. In capital cases, juries can also have the responsibility of deciding the punishment. See *grand jury, petit jury.*

jury consultant   a consultant whose specialty is assisting lawyers in selecting and addressing jurors in ways that improve their odds of returning a favorable *verdict.* Some jury consultants are psychologists whereas others come from other backgrounds.

jury deliberations   the discussions by jurors in a criminal case that eventually lead to a *verdict* or to a *hung jury.*

jury duty   the practice and obligation of citizens serving on juries for local courts. Those chosen for jury duty generally are selected from the rolls of registered voters.

jury instructions   verbal instructions given by a trial judge to members of a *petit jury* prior to commencement of the trial. See *jury nullification.*

jury nullification   the disregard by a jury of the *evidence* presented and the rendering of its *verdict* based on other criteria.

jury selection   the process of picking by elimination jurors for a trial by the prosecutor and defense counsel. In highly publicized cases, defense counsel may use the services of a *jury consultant.* See *peremptory challenge, voir dire.*

jury system   a system of justice employing a body composed of the accused's peers who allegedly can hear the facts and render an impartial verdict.

jury tampering   a criminal offense revolving around an attempt to threaten, bribe, or otherwise influence one or more members of an impaneled *jury.* Jury tampering is most closely associated with *organized crime* which has a history of trying to influence jurors in order to control the outcome of a criminal trial.

J

jury trial a criminal trial making use of a *jury* to hear facts and decide the guilt of the *defendant*. Compare with *bench trial*.

just conviction a criminal conviction where the defendant is correctly found guilty of the crime. Compare with *wrongful conviction*.

just deserts the principle that asserts that the punishment for a crime should be commensurate with the seriousness of the crime; you get what you deserve.

justice what is considered just and fair. Also, a *judge* of a supreme court.

justice model a philosophy and its associated policy that emerged in the wake of the 1970s *nothing works* movement. The justice model emphasized predictability of legal consequences for offenders.

justice of the peace a local official who has the authority to issue licenses, perform marriages, and adjudicate traffic and other minor offenses. Compare with *judge, magistrate*.

Justice Policy Center a research center within the *Urban Institute* that conducts research and evaluation to improve criminal justice.

Justice Reinvestment Initiative a collaboration of the *Bureau of Justice Assistance* and *Pew Charitable Trusts* to promote a data-based approach to increase public safety and reduce spending on criminal justice.

Justice Research and Statistics Association (JRSA) a national organization whose membership primarily consists of *statistical analysis center (SAC)* staff. Among the services JRSA offers are specialized training in criminal justice data analysis techniques. The organization holds annual meetings and is headquartered in Washington, D.C.

Justice Studies Association (JSA) an international organization of scholars, practitioners, activists, and others committed to promoting transformative justice. The JSA is headquartered at Bridgewater State University in Massachusetts and publishes the journal *Contemporary Justice Review*.

justifiable homicide the taking of a life that is excusable due to self-defense or some other consideration.

Just Say No an anti-drug campaign of the 1980s which rested on the premise that youths and other potential drug users would refrain simply by the internalization of this message. Also, the slogan of this campaign. First Lady Nancy Reagan was one of the campaign's most vocal proponents.

juvenile one who is under the age of consent, which is most often eighteen. Compare with *adult*.

juvenile bind over the transfer of a juvenile case to adult court. Juvenile bind over is used in serious offenses, such as *homicide, rape, robbery*, and serious assault. Once bound over, the juvenile's case is handled like any other felony case.

juvenile court   a court of law, sometimes a subdivision of a common pleas court or domestic relations court, which has jurisdiction over matters pertaining to the delinquency and unruliness of persons who have not yet attained adulthood.

juvenile delinquency   criminal offenses committed by those under the age of legal responsibility, most often 18 years of age.

juvenile delinquent   term describing a youth who has engaged in *juvenile delinquency*.

juvenile detention center   a temporary holding facility for youthful offenders. Also referred to as *attention center*. Same as *detention center*.

juvenile gang   a *gang* whose members are predominantly made up of juveniles.

juvenile homicide   homicide committed by a *juvenile offender*.

juvenile justice   the entire system designed to process and meet the needs of offenders under the age of 18.

Juvenile Mentoring Program (JUMP)   a grant program of the *Office of Juvenile Justice and Delinquency Prevention (OJJDP)* designed to promote mentoring in order to specifically address poor school performance and the dropout rate in schools.

juvenile offender   an *offender* under the age of 18.

juvenile sex offender   a *juvenile offender* who has committed one or more sex offenses.

juvenile transfer   see *juvenile bind over*.

K-9 a unit in law enforcement agencies that employs trained dogs to assist with locating and apprehending criminal suspects. Compare with *cadaver dog*.

Kaczynski, Theodore (1942–present) a former university professor who was convicted of several fatal bombings in the United States. Kaczynski was a brilliant mathematician who had studied at Harvard and once held an academic post at the University of California at Berkeley. He was known as the *Unabomber*. See *domestic terrorism*.

Kallikaks chronicled by Henry Goddard, a prominent eugenicist, a family that spawned a number of feebleminded and criminal offspring over the course of several generations. Based on his research, Goddard concluded that crime is inherited. See *Jukes family*.

kangaroo court an unofficial court that is often held by a mob or *vigilante* group whose judgment is biased against the *accused* and rendered without *due process*.

Kansas City Massacre an infamous shootout in 1933 in Kansas City, Missouri between law enforcement and criminals which left four officers dead.

Kansas City Preventive Patrol Experiment a progressive law enforcement experiment of the 1970s where police were assigned to patrol certain high crime areas in an attempt to reduce crime.

Karpis, Alvin (1907–1979) a gangster and member of the Barker-Karpis gang that was active in the 1930s. Karpis was caught, convicted, and sent to *Alcatraz*.

Keating, Charles (1923–2014) a banker and real estate developer who played a role in the savings and loan scandal of the late 1980s. See *white-collar crime*.

Kefauver Committee a committee convened by Senator Estes Kefauver in the 1950s to investigate *organized crime* in the United States. Compare with *McClellan Committee*.

K

Kelly, George "Machine Gun" (1895–1954)   a *gangster* of the 1930s who was known for using a *Thompson submachine gun*. He was caught, convicted, and sentenced to prison.

Kerner Commission   shortened name for the National Advisory Commission on Civil Disorders. The Kerner Commission, chaired by Senator Otto Kerner, was a federal response to the racial unrest of the 1960s. The Kerner Commission issued its report on urban race riots in 1968.

ketamine   a drug used mainly as an analgesic and for anesthesia that is sometimes abused for recreational purposes. Ketamine use can result in death. See *substance abuse*.

Kevlar   a synthetic material used to make *body armor*. Despite its protective attributes, Kevlar can be penetrated by Teflon-coated bullets or those designed to pierce armor plate. See *bulletproof vest, cop-killer bullets*.

Kevorkian, Jack (1928—2011)   a Michigan pathologist who gained notoriety by assisting elderly and terminally ill patients with taking their own lives. Kevorkian was tried multiple times for murder and eventually convicted. See *assisted suicide*.

khaki-collar crime   a crime occurring within the military. Compare with *camouflage-collar crime, white-collar crime*.

kickback   a payment to an official by a contractor for favoritism in awarding the contract or any payment exacted as a condition for granting assistance.

kidnapping   the willful deprivation of another person's liberty, usually in order to secure ransom. Perhaps the most famous kidnaping of the 20th century was the *Lindbergh kidnapping* in 1932. Compare with *abduction*.

kill   to bring about the cessation of life. See *murder.*

kingpin   the head of a narcotics trafficking organization. Also referred to as a *drug kingpin*. Compare with *boss*.

King, Rodney (1965–2012)   a Los Angeles motorist subjected to a severe beating at the hands of Los Angeles police officers in 1991 which was videotaped. King was stopped for a routine traffic arrest, after which he was beaten by several officers. Several of the officers were later found guilty of *excessive force*, resulting in suspensions, firings, and criminal convictions. The case led to the formation of a commission to investigate the Los Angeles Police Department. See *police brutality*.

KlaasKids Foundation   named for young homicide victim *Polly Klaas*, a foundation whose mission is to stop crimes against children.

Klaas, Polly (1981–1993)   a 12-year-old girl who was abducted from her home in Petaluma, California and murdered by *Richard Allen Davis*.

Klan Watch former name of *The Intelligence Project.* See *Ku Klux Klan.*

Klebold, Dylan (1981–1999) one of the two perpetrators of the *Columbine massacre.* Klebold took his own life.

kleptomania a psychological compulsion to steal.

Knapp Commission the commission convened in the early 1970s to investigate widespread corruption within the New York City Police Department. Officer *Frank Serpico* was instrumental in bringing this corruption to public light, resulting in numerous indictments and convictions of police officers. See *grass-eater, meat-eater, police corruption.*

Knox, Amanda (1987–present) an American woman who, along with a male *codefendant*, was convicted of killing an English student in Italy. After a sensational trial, Knox was sentenced to 28 years in prison. She was jailed for four years, appealed, and was released in 2011. She was subsequently found guilty again by the Italian courts in 2014.

Kornhauser, Ruth (1926–1995) a sociologist who conducted pioneering research on *juvenile delinquency.* Kornhauser earned her Ph.D. in sociology at the University of Chicago.

Kretschmer, Ernst (1988–1964) a German psychiatrist who devised a typology of body types and associated them with personality traits, including some related to criminal behavior. See *constitutional theory.*

Ku Klux Klan a racist, white supremacist organization, found mostly in the southern United States, dedicated to asserting the superiority of whites over racial, ethnic, and religious minorities, including Blacks, Jews, and gays. See *The Intelligence Project.*

labeling the processing of offenders and other deviants when described or designated with a new identity, which in turn can have negative consequences.

labeling perspective a theoretical perspective in the sociology of deviant behavior which holds that agents of social control label those whom they process, sometimes leading to limited opportunities and further deviance. Also referred to as labeling theory. See *primary deviance, secondary deviance, stigma.*

labeling theory see *labeling perspective.*

labor camp a camp where sentenced offenders are required to perform physical labor. Compare with *gulag.*

Lacassagne, Alexandre (1843–1924) a French physician and criminologist who founded the *Lacassagne School of Criminology.*

Lacassagne School of Criminology a school of thought led by *Alexandre Lacassagne* that argued for the importance of both social and individual factors in the genesis of crime.

La Cosa Nostra another term for the *Mafia.* In his testimony before the *McClellan Committee* hearings in the 1960s, reportedly *Joseph Valachi* was the first to use this term to describe the Sicilian underworld. See *organized crime.*

lands the ridges between the rifling *grooves* in the barrel of a firearm. Lands, like grooves, leave distinct marks on bullets passing through the barrels of rifled firearms making possible matches between bullet and firearm. See *ballistics.*

Lansky, Meyer (1902–1983) a major figure in *organized crime* in the mid-20th century.

Lanza, Adam (1992–2012) young man responsible for the *Sandy Hook Massacre,* also known as the Newtown Massacre. Lanza shot and killed his mother before going to the school, and he took his own life as police were closing in.

larceny   same as *theft*.

latent fingerprint   see latent prints.

latent fingerprint examiner   a fingerprint examiner who specializes in the analysis of *latent prints*.

latent function   unintended or unanticipated positive consequences of laws or rules. Compare with *manifest function*.

latent prints   fingerprints lifted from surfaces other than the hands bearing the prints. See *arch, fingerprint, whorl*.

Latin Kings   a notorious *gang*. Compare with *Bloods*, *Crips*, *Gangster Disciples*.

Law and Society Association (LSA)   a scholarly organization made up of socio-legal scholars who study the influence of law on social relations. The LSA holds its annual meeting in August in conjunction with the meetings of the American Sociological Association. See *socio-legal studies*.

Law Enforcement Assistance Administration (LEAA)   a large agency in the U.S. Department of Justice created by the *Omnibus Crime Control and Safe Streets Act of 1968*. The LEAA distributed millions of dollars for various crime control programs, some of which drew criticism for their waste of funds on such purchases as armored personnel carriers for small-town police departments.

law-related education   a large-scale coordinated effort to lessen youth violence and delinquency by involving youths in learning about the law and developing alternatives to *violence, substance abuse*, and other maladaptive behaviors. See *Youth for Justice*.

learning theory   any one of several theories in criminology which asserts that offenders learn criminal behavior in association with other offenders through principles of operant conditioning or other methods. An example is *differential association*.

Lee, Henry C. (1938–present)   a well-known criminalist and former professor at the University of New Haven. Lee consulted in a number of highly publicized criminal cases, including those of *O. J Simpson* and *JonBenet Ramsey*.

left realism   a school of thought in criminology that recognizes the need to control crime, which is disproportionately rooted in social disadvantage and deprivation.

legal aid attorney   an attorney that represents indigent clients most often in civil matters. Compare with *court appointed attorney, public defender*.

legal aid society   a private, nonprofit organization of attorneys that exists to represent the rights of indigent clients. See *appointed counsel, public defender.*

legal dose   the amount of drug recommended to treat the condition in question. Compare with *lethal dose.*

legal factors   factors, such as an individual's culpability and the nature and extent of the charges that appropriately should be considered in *prosecution* and in court. Also known as legal variables. Compare with *extralegal factors.*

legalization   the process of making legal that which was previously illegal. Compare with *decriminalization.*

legal limit   the blood alcohol level beyond which a person is considered legally intoxicated. See *blood alcohol content, drunk driving.*

legal medicine   the field that focuses on the interface of law and medicine and medical practice.

leg irons   *manacles* designed for use on the legs. Leg irons, especially those used in the past, were heavy and constrained the person's mobility.

Leonard, Elmore (1925–2013)   a novelist and screenwriter who wrote numerous crime novels, some of which became motion pictures. See *crime novel.*

Lepine, Marc (1964–1989)   a Canadian mass murderer who shot and killed 14 students at the University of Montreal in 1989. Lepine took his own life. See *mass murder.*

less-than-lethal force   force used by law enforcement to subdue criminal suspects which does not result in loss of life. Examples of less-than-lethal force include *pepper spray* and the *Taser.*

lethal dose   the amount of a drug or other substance sufficient to bring about death. Lethal dose can vary by individuals. See *lethality.*

lethal force   force used by law enforcement, prison guards, and other criminal justice security officials which could result in the death of suspects or inmates. Compare with *non-lethal weapon.*

lethal gas   the combination of cyanide and sulfuric acid used to execute condemned prisoners. See *gas chamber, death penalty.*

lethal injection   a form of *capital punishment* in which lethal amounts of various prescription drugs are administered to the condemned until life ceases. Despite its popularity, lethal injection can result in a *botched execution*. Compare with *electric chair, firing squad, hanging.*

lethality   the extent to which something is likely to cause death. The lethality of a firearm is greater than that of a human fist.

lethal violence   violence that results in death.

Letourneau, Mary K. (1962–present)   a schoolteacher who had a sexual relationship with a 12-year-old student. She was convicted of raping the student and served six years in prison after violating terms of a plea agreement. Letourneau and the student eventually married. See *statutory rape*.

Level of Service Inventory-Revised (LSI-R)   an instrument based on offender attributes and circumstances used to predict parole outcome, institutional misconduct, and recidivism. The LSI-R quantitative survey is proprietary and a user must be trained and certified in its use.

level-pulling strategy   a coordinated approach to *crime control* that involves a wide variety of community actors, including law enforcement, justice officials, and social service providers.

lex talionis   the principle in criminal law which embodies the notion of retaliation. *Lex talionis* implies that society's response to criminal offenses will be commensurate with the seriousness of the offense and the harm done. Compare with *retribution*.

libel   a false published statement that damages the name or reputation of another. Compare with *slander*.

lie detector   a device using instruments to register changes in blood pressure, strength of pulse beat, respiratory movements, or increased perspiration as indicative of lying under questioning. See *polygraph test*.

life imprisonment   same as *life sentence*.

lifer   an inmate who has been given a *life sentence*.

life sentence   a judicially imposed term of confinement equal to the life of the sentenced offender. Offenders who receive a life sentence will die in prison unless they are released early through a mechanism such as a *pardon* or *parole*.

LifeSkills Training   a prevention program targeted toward middle school students that addresses a wide variety of social, psychological, and other factors associated with *substance abuse* and *violence*. LifeSkills Training is part of the *Blueprints for Healthy Youth Development*.

life without parole   a *life sentence* in a penal institution which eliminates the possibility of release by parole authorities. Life without parole practically means that the convicted offender will die in prison.

lifting   the process of transferring and securing fingerprints that are left at a *crime scene*.

ligature strangulation   strangulation involving the use of a rope, cord, or other flexible medium. See *strangling*.

Lindbergh kidnapping   the 1932 *kidnapping* and *murder* of the son of famed aviator Charles Lindbergh.

lineup   an array of persons for the purposes of suspect identification. A lineup consists not only of the suspect, but also other individuals, which sometimes includes police officers in street clothes. The victim or witness tries to identify the suspect from this array of people. See *eyewitness identification*. Also known as *identification parade*. Compare with *showup*.

linkage blindness   the inability of law enforcement officials to make connections between related crimes. Linkage blindness is a problem in serial crimes, such as homicide or sex offenses, where similar offenses occur in multiple jurisdictions. Greater participation in programs like *VICAP* (Violent Criminal Apprehension Program) might reduce or eliminate linkage blindness.

List, John (1925–2008)   an unemployed accountant who in 1971 murdered his mother, his wife, and four children and then fled. List was identified by a viewer of an *America's Most Wanted* when his case was featured on the television show. He was subsequently found guilty and sentenced to serve six consecutive life sentences. List's capture was facilitated by *forensic art*.

littering   most often a misdemeanor, the offense of strewing or discarding trash or other material in public.

LiveScan   a patented, electronic system for scanning human fingerprints. Based on complex mathematical algorithms, LiveScan captures the precise contours of the *fingerprint*, preserving it for later comparison with other prints.

lividity   the collection of blood in a corpse caused by gravity. The presence of lividity can be used to determine if a body has been moved after the blood has settled.

loan sharking   the lending of money at rates of interest far above what is considered reasonable. Generally those who borrow under these circumstances have a very limited time to pay both the principal and the interest. Failure to pay can result in serious injury or even death. Compare with *usury*.

lobotomy   a psychosurgical procedure in which a cut is made into the frontal lobe of the brain, rendering the patient less violent and more docile. Once a staple in psychiatric treatment, the procedure has fallen into disuse. See *psychosurgery*.

Locard, Edmund (1877–1966)   an early forensic scientist who is best known for *Locard's Exchange Principle*.

L

Locard's Exchange Principle   the notion that criminals both leave material at a crime scene and take away something from it, all of which has potential forensic value. See *Locard, Edmund.*

lockdown   a procedure in a prison where officials confine all prisoners in their cells. Prison officials use lockdowns when they want to confiscate *contraband* or they believe a *prison riot* is imminent.

lockup   a holding facility for offenders, often located within a police station, designed for short-term detention. Compare with *jail, workhouse.*

logic model   a chart showing the inputs, activities, and outcomes of a program for the purpose of evaluation. See *program evaluation.*

loitering   a petty offense characterized by an individual's presence in a public or private place without official business or other reasonable justification. Loitering is a *public order crime.*

LoJack   a brand name of an electronic device installed in a motor vehicle that enables authorities to locate and track it if it is later lost or stolen. Despite their role in recovering stolen vehicles, LoJacks and similar devices are controversial because of the "big brother" aspect of knowing the motorist's whereabouts at any given time.

Lombroso, Cesare (1835–1909)   an Italian physician and criminologist of the early 20th century who asserted that criminals were throwbacks to an earlier form of human. Lombroso's contributions to the study of crime included an emphasis on the systematic collection of data. See *atavism, biocriminology.*

long gun   a firearm with a stock and long barrel, such as a *rifle* or *shotgun.* Compare with *handgun.*

longitudinal study   a study of youthful or other offenders where personal and legal data are periodically collected at various points along the life course. Longitudinal studies are expensive and take a long time to yield definitive results. Other problems with this approach include trying to stay in touch with subjects who periodically move and whose memories may not serve them well when asked by researchers to recall certain events. Notable examples of longitudinal studies are the National Youth Study administered by the University of Colorado, the Pittsburgh Youth Study administered by the University of Pittsburgh, the Rochester Youth Study administered by the University at Albany, and the *Program on Human Development in Chicago Neighborhoods (PHDCN).*

lookout   a person who keeps watch as other offenders engage in criminal conduct. Compare with *codefendant.*

loop   a loop-shaped part of a human *fingerprint*. Compare with *arch, whorl*.

loot   money or other ill-gotten goods.

looting   the indiscriminate *theft* of goods from residential or commercial establishments often during a riot or in the aftermath of a natural disaster, such as a tornado or hurricane.

loss prevention   a set of practices employed by companies to preserve profit by preventing *employee theft*.

low explosive   explosive materials that burn rather than explode. Compare with *high explosive*.

LSD   abbreviation for lysergic acid diethylamide. Commonly called acid. A hallucinogenic drug highly favored by the youthful counterculture in the 1960s, it was popularized in song by many psychedelic recording artists and was the drug of choice of drug guru and pop icon, Timothy Leary. See *hallucinogen, substance abuse.*

Lucas, Henry Lee (1936–2001)   an American *serial killer* who was responsible for numerous murders in several states. He committed some of his crimes with *Ottis Toole*. See *serial murder.*

Luciano, Charles "Lucky" (1897–1962)   an American *organized crime* figure of Italian ancestry and first *boss* of the Genovese *crime family*. See *Lansky, Meyer.*

Luminol   a chemical that reacts to iron found in hemoglobin that is used to detect the presence of blood at crime scenes. Luminol fluoresces under an ultraviolet light.

lush worker   one who steals wallets from drunks, often by cutting their pockets. Compare with *cutpurse, jack roller, pickpocket.*

lynching   the practice of illegally taking the life of another by hanging, generally accomplished by a mob and often motivated by racial or ethnic hatred or a desire for immediate *retribution*. Now infrequent, lynching in the United States is associated with white supremacists and their targeting of Blacks. See *hate crime, Ku Klux Klan.*

MacDonald, Jeffrey (1943–present) an Army Green Beret and military physician who was convicted of murdering his wife and two daughters in 1970. MacDonald maintains that a cult of hippies carried out the murders.

mace an aerosol chemical designed to render an attacker temporarily immobile and harmless. Compare with *pepper spray*. See *non-lethal weapon*.

macro-level criminological variables that are measured at the structural level, such as poverty and population change. Compare with *microlevel*.

MADD abbreviation for *Mothers Against Drunk Driving*.

Madoff, Bernard (1938–present) a stockbroker and financier based in New York who was convicted of perpetrating a large *Ponzi scheme*.

Mafia a secret society of *organized crime* of Italian or Sicilian origin. See *crime family*.

magazine the often detachable container that holds cartridges for a firearm. The capacity of magazines is the subject of debate and legislation with the rationale that higher capacity magazines increase the *lethality* of firearms. Also referred to as a clip.

magistrate a quasi-judicial who presides over minor cases and preliminary hearings. Compare with *judge*.

MAGLOCLEN abbreviation for the Middle Atlantic-Great Lakes Organized Crime Law Enforcement Network. See *Regional Information Sharing System (RISS)*.

mail-order fraud fraud perpetrated by use of the mail. Examples of mail-order fraud include the sale of diplomas through the mail.

major case squad a unit within some law enforcement agencies that focuses on unusually serious crimes.

male prostitute a man or boy who exchanges sexual acts for money, drugs, or other forms of currency.

malfeasance the intentional act of engaging in misconduct by a person in the performance of their official duties, generally in the public service. Compare with *misfeasance, nonfeasance.*

malice aforethought an individual's predetermination to commit an unlawful act, or the intent to do wrong that precedes an offense.

malicious shrinkage the reduction of inventory as a result of *employee theft.* Compare with *pilfering.*

malignant narcissism a form of *narcissistic personality disorder* that includes *antisocial personality disorder, aggression,* and *sadism.* Compare with *The Dark Triad.*

malum in se the Latin term meaning bad in and of itself. In the history of criminal law, some offenses, such as incest and cannibalism, were deemed so reprehensible that they were considered crimes inherently wrong irrespective of any laws prohibiting them. Examples of offenses *mala in se* include homicide, rape, and robbery. In most societies such offenses are subject to severe criminal penalties. Compare with *malum prohibitum.*

malum prohibitum the Latin term meaning bad because it has been prohibited. An example of *malum prohibitum* is drug abuse, which historically and universally has not been deemed inherently wrong. Compare with *malum in se.*

Malvo, Lee (1985–present) a Jamaican-American who, along with *John Allen Muhammad,* perpetrated the *Beltway Sniper* shootings in 2002. He was sentenced to *life without parole.* See *sniper.*

manacles restraining devices consisting of heavy metal rings designed to link a prisoner's wrists or ankles. Compare with *handcuffs, leg irons.*

mandatory arrest a policy in which those accused of domestic violence are automatically arrested. Mandatory arrest policies resulted from the realization that victims of domestic violence remain vulnerable unless the assailant is arrested and jailed.

mandatory minimum a minimum term of confinement prescribed by law for persons convicted of certain offenses.

mandatory release the release of a convicted offender from a *correctional facility* due to *good time credit* or other statutory provisions.

mandatory sentence a criminal sentence which must be served. Mandatory sentences became popular in the 1970s and 1980s with the abandonment of the rehabilitative ideal. Since inmates must serve more time than under *indeterminate sentencing,* mandatory sentences cause prison populations to burgeon. See *indeterminate sentence, nothing works, rehabilitation.*

Manhattan Bail Project an early large-scale experiment in releasing criminal defendants on their own recognizance instead of having them post cash bail. See *pretrial release*.

manhunt an organized search by law enforcement, correctional or other officials for an escaped or otherwise wanted *offender*. Manhunts can span counties, states, regions, or entire countries. The manhunt for *Osama bin Laden* in the aftermath of the *Attack on America* was perhaps one of the biggest in history.

man hunter one who tracks down criminals, particularly violent ones.

manic-depression a mental disorder where the individual experiences wide mood swings between elevated happiness and depression. The term bipolar disorder has replaced manic depression as the preferred terminology.

manifest function the intended or anticipated consequences of law or rules. A manifest function of a law increasing the penalty for armed robbery is to deter would-be robbers from committing such crimes. Compare with *latent function*.

Mann Act a federal act passed in 1910 that prohibits the interstate transport of women and girls for the purpose of *prostitution*. See *human trafficking*.

manner of death the way in which a person dies. The manner of death differs from the actual *cause of death*. For example, where the actual cause of death might be listed as cerebral anoxia, the manner of death is hanging.

Mannheim, Hermann (1889–1974) a well-known English *criminologist* of the 20th century. Mannheim was affiliated with the London School of Economics. His students included the late criminologist and law professor *Norval Morris*.

Mano Nera same as *Black Hand*.

manslaughter the non-intentional taking of a human life. Manslaughter is considered less serious than *murder*. See *involuntary manslaughter, voluntary manslaughter*.

Manson, Charles (1934–present) leader of the so-called Manson Family who was responsible for the 1969 murders of actress Sharon Tate and four guests at her home. The Manson Family later murdered Leno and Rosemary Labianca. Manson was sentenced to death, but his sentence was later commuted to a life sentence. Every time Manson has come up for parole, his petition is denied, not only because of the gravity of his crimes, but also due to the efforts of prosecutors and the victims' families. See *mass murder*.

manual strangulation the strangulation of an individual using one's hands. Compare with *ligature strangulation*.

M

marbling   an effect on human body produced by the hemolysis of blood vessels in reaction to hemoglobin and hydrogen sulfide. Marbling takes its name from the resultant green, purple, and red discoloration of the skin.

marijuana   cannibis sativa, an organic hallucinogenic drug. See *substance abuse*.

Marion   the site of a U.S. Penitentiary at Marion, Illinois. The facility at Marion has the reputation for housing some of the most dangerous federal convicts.

marital rape   sexual intercourse forced on one spouse by the other.

mariticide   the killing of one's husband. Compare with *uxoricide*.

mark   the intended victim in a *confidence game* or swindle.

market reduction approach   crime control efforts aimed at curbing theft by increasing penalties for those who knowingly buy stolen goods. See *fence*, *receiving stolen property*.

marshal   a law enforcement officer associated with a judicial district.

Marshall hypothesis   hypothesis derived from Justice Thurgood Marshall's opinion in the Furman decision, which asserts that most people have little knowledge about the death penalty, but if they did, they likely would oppose it.

Martin, Trayvon (1995–2012)   a 17-year-old African American youth who was shot and killed by *George Zimmerman* in Sanford, Florida. See *stand your ground laws*.

Marxist criminology   a theoretical perspective in criminology grounded in the writings of Karl Marx. Compare with *conflict model*.

mask of sanity   expression coined by psychologist Hervey Cleckley to describe the seemingly normal persona of the *psychopath*. See *psychopath*, *antisocial personality disorder*.

masochism   the practice of deriving sexual pleasure from having pain inflicted. Compare with *sadism, sadomasochism*.

Massachusetts Treatment Center   a well-known *sex offender* treatment facility housed within the Bridgewater Correctional Complex in Bridge-water, Massachusetts.

massacre   the murder of many innocent persons in a single incident. Compare with *mass murder*.

massage parlor   a business ostensibly offering massages, but which is a front for *prostitution* services.

mass incarceration   the confinement in jails or prisons of large numbers of people, particularly people of color.

mass murder   the killing of four or more people in the same location and during the same event. Usually committed by an individual, but there can be multiple perpetrators. Do not confuse with *serial murder*. See *Huberty, James, Lanza, Adam, Lepine, Marc, Speck, Richard, and Whitman, Charles*. Compare with *serial murder, multiple homicide, spree murder*.

mass spectrometer   a machine used in forensic laboratories to identify chemical composition of substances. Mass spectrometers can be used to analyze paint and other samples in the investigation of crimes.

masturbation   sexual gratification by self-stimulation of one's genitals.

maternal incarceration   the confinement in jail or prison of mothers who have dependent children. Maternal incarceration has profound negative consequences on families, including frequent change of locations and schools, and separation from support networks.

matricide   the killing of a mother by one or more of her children.

maturation   the process over time resulting in an *offender* who desists from engaging in crime. See *desistance*.

maximum security   a level of security in correctional facilities for inmates who are assaultive or otherwise dangerous, or who pose an escape risk. Compare with *close security, medium security, minimum security, trusty security*.

maximum security prison   a penal institution which houses the most violent offenders. Compare with *supermax prison*.

maximum sentence   the most severe sentence for any given crime. Compare with *minimum sentence*.

may issue   expression used to describe firearms laws which give officials discretion in who receives a license to carry firearms. Compare with *shall issue*.

mayor's court   a local court presided over by a mayor of a town or municipality that hears minor criminal and traffic cases, such as *driving under the influence* and speeding.

McClellan Committee   a U.S. Senate committee convened to investigate *organized crime* in the late 1950s.

McCord, Joan (1930–2004)   a prominent criminologist who was an early advocate of longitudinal research on crime. McCord was the first female president of the *American Society of Criminology*. See *longitudinal study*.

McDonald Laurier Institute (MLI)   a Canadian conservative think tank that has issued papers on criminal justice issues, including prison radicalization. The MLI is headquartered in Ottawa, Canada.

McKay, Henry D. (1899–1980)   a criminologist of the early to mid-20th century who was one of the pioneers in the ecological study of crime and social disorder. McKay was affiliated with the *Institute for Juvenile Research*. See *Chicago Area Project, Chicago School of Criminology.*

McVeigh, Timothy (1968–2001)   a former Army veteran who with *Terry Nichols* perpetrated the *Oklahoma City bombing*. McVeigh was convicted of 160 counts of murder and was executed by *lethal injection*. See *domestic terrorism.*

MDPV   a potent, synthetic euphoric stimulant used as a substance of abuse. See *substance abuse.*

means   the ability to carry out a crime. Compare with *motive* and *opportunity.*

meat-eater   a corrupt law enforcement officer who aggressively pursues opportunities for graft. Compare with *grass-eater*. See *Knapp Commission, police corruption, Serpico, Frank.*

mechanical theory of crime and incarceration   the theory that changes in the *crime rate* bring about changes in the *prison population.*

mechanism of death   the actual physiological changes that take place in the human organism to bring about death. Compare with *cause of death, manner of death.*

Medellin cartel   a powerful narcotics *trafficking* organization based in Medellin, Colombia. Compare with *Cali cartel.*

mediation   the practice of bringing together adversaries for the purpose of discussing and reconciling differences. See *victim-offender mediation.*

media violence   violent acts portrayed in various print and electronic media.

Medicaid fraud   fraud perpetrated by physicians, pharmacists, or insurance companies by over-billing or false billing for services. See *corporate crime, white-collar crime.*

medical examiner   a medically trained doctor authorized to perform official examinations of deceased individuals to determine *cause of death, mechanism of death* and *manner of death*. Compare with *coroner.*

medical marijuana   marijuana used to alleviate the discomfort or other symptoms of a disease, usually cancer. Medical marijuana has been surrounded by controversy because of the drug's illegal status in most jurisdictions.

Medicare fraud   The dishonest acquisition of Medicare benefits by citizens or Medicare reimbursements by medical practitioners or health organizations.

medicolegal   having to do with the interface between medicine and law.

medium security   security level for correctional facilities between minimum and maximum security for inmates who pose few problems, but who still require supervision. Compare with *close security, maximum security, minimum security, supermax prison, trusty security.*

Meese Commission   a federal commission led by U.S. Attorney General Edwin Meese in the 1980s to investigate *pornography* in the United States.

Megan's law   informal name for the Sex Offender Registration Act, a federal law requiring convicted sex offenders to register with local law enforcement on their release from prison. Megan's law was named for Megan Kanka, a New Jersey girl who was murdered by a convicted *sex offender* who lived across the street from her. See *sex offender notification, sex offender registration.*

Memphis Model   a model for a *crisis intervention team* that involves training law enforcement to recognize the unique challenges posed by those suffering from mental illness.

Menendez brothers, Joseph (1968–present) and Erik (1970–present)   two brothers who murdered their affluent parents in California. The Menendez brothers maintained in their defense that they had been subjected to extreme physical and emotional abuse, primarily by their father. Their trial, which highlighted on *Court TV*, ended with their convictions. They were sentenced to life in prison. See *parricide.*

mens rea   a Latin term meaning guilty mind or criminal *intent*. Mens rea traditionally is necessary in order for an act to be considered criminal. See *actus reus.*

mental health court   a *specialty court* which is designed to ensure the appropriate treatment of defendants suffering from mental illness. Pioneered in Florida and Washington, mental health courts followed the model set by *drug courts.*

mentally ill offender   a person suspected, charged, or convicted of a crime who suffers from some form of mental illness. Mentally ill offenders pose a special challenge to the criminal justice system that is ill-prepared to diagnose and treat them. See *crisis intervention team, Memphis model, mental health court.*

mercy   compassion toward offenders.

mercy killing   the intentional killing of a person for the purpose of alleviating suffering. An example of mercy killing is the elderly man who shoots his terminally ill wife so she will not have to suffer from a painful or terminal condition. See *Kevorkian, Jack*. Compare with *assisted suicide*.

Merton, Robert K. (1910–2003)   an American sociologist noted for his oftcited article, "Social Structure and Anomie," in which he identified five ends to which individuals aspire and the deviant means they sometimes use to achieve them. Merton spent most of his career at Columbia University. See *anomie theory, strain theory*.

mescaline   a *hallucinogen* made from the peyote cactus. See *substance abuse*.

mesomorph   a body type characterized by muscularity. Mesomorphs were once thought to be aggressive criminals. See *Sheldon, William*.

meta-analysis   a methodology used to synthesize the results of a large number of evaluative studies. Meta-analysis involves an examination of the statistical conclusions of the studies' results, permitting the analyst to integrate their findings. Applications of meta-analysis in criminal justice include the analysis of treatment literature in corrections.

methadone   a synthetic form of narcotic used as a substitute for *heroin* in the treatment of addiction. Methadone does not have the same withdrawal effects as heroin.

methadone diversion   a program designed to divert heroin-addicted offenders from regular criminal justice processing. If the offender completes the program as prescribed, the criminal charges will be dropped. See *diversion*.

methadone maintenance   a program for those addicted to heroin that permits them to take the drug methadone under a physician's care in order to eliminate the severe symptoms associated with withdrawal from heroin. Some critics object to such programs because they believe that, in essence, it is substituting one drug, albeit a legal one, for another.

methamphetamine   a synthetic derivative of amphetamine. Originally manufactured and controlled by outlaw motorcycle gangs, methamphetamine can be produced in labs small enough to fit into a motor vehicle. The effects of methamphetamine include increased alertness, euphoria, and decreased appetite. See *substance abuse*.

meth lab   a *clandestine drug lab* that manufactures *methamphetamine*. The chemicals used in meth labs pose an *environmental hazard*.

method of operation   the specific techniques used by criminals to perpetrate their offenses. Also called *modus operandi*.

microbial biosecurity threat agent   infectious microorganisms that can be used by terrorists to inflict mass atrocities. See *bioterrorism*.

microlevel criminological variables which are measured and analyzed at the individual level. Compare with *macro-level*.

midnight basketball a delinquency prevention program mentioned in the 1994 Crime Bill to curb inner-city crime in the United States by keeping urban youths off the streets and engaging them with alternatives. Midnight basketball, which was promoted with good intentions, was ridiculed by critics.

Milgram Study named for psychologist Stanley Milgram, research studies that explored *obedience to authority,* such as that of the Nazis during the *Holocaust.*

military crime crime committed by military personnel. See *khaki-collar crime*. Compare with *war crimes*.

military justice the system of justice in any branch of the Armed Forces. See *Uniform Code of Military Justice.*

military police law enforcement authorities of any branch of the armed forces. Military police function in much the same way as their civilian counterparts. Often referred to as MPs.

military prison a prison operated by a branch of the government in which prisoners of war and those convicted of military crimes are incarcerated. See *Abu Ghraib.*

Milken, Michael (1946–present) an infamous insider trader of the 1980s. Milken, along with *Ivan Boesky* and others, was convicted and sentenced to serve ten years' imprisonment, a stiff sentence for this type of crime. His sentence was later reduced for his cooperation in other criminal cases. See *insider trading, white-collar crime.*

mind hunter a term used to describe a *profiler* who must probe the psychological motives of the *offender* in order to solve a crime. See *criminal investigative analysis, serial murder.*

minimum security a level of security in correctional facilities for inmates who pose the least amount of risk for escape or rule infraction. Compare with *close security, maximum security, medium security, trusty security.*

minimum sentence the least amount of time an offender faces for a particular crime. Compare with *maximum sentence.*

Ministry of Justice a government agency in some countries responsible for overseeing a variety of justice-related activities.

minority overrepresentation the disproportionate number of minorities in criminal justice populations. Compare with *disproportionate minority confinement.*

M

minor offense   an offense which represents little or no harm to society.

Miranda warning   a warning law enforcement officers are required to give to arrested felons. Suspects must be advised that they have the right to remain silent; that any statement they make can be used against them in a court of law; that they have the right to an attorney, and if they cannot afford an attorney one will be appointed before any questioning. The Miranda warning stems from the 1966 U.S. Supreme Court case *Miranda v. Arizona*.

miscarriage of justice   the failure of the criminal justice system to serve the ends of justice. An example of a miscarriage of justice is when an innocent person is found guilty and sentenced to prison. See *convicted innocent, wrongful conviction*.

misconduct in office   any conduct in the performance of official duties that constitutes criminal behavior or a violation of ethical standards. Misconduct can include *malfeasance, misfeasance*, or *non-feasance*.

misdemeanor   a minor crime generally punishable by a fine or a term of confinement of less than one year. Misdemeanors typically are adjudicated in municipal, county, or other lower courts having limited jurisdiction. Compare with *minor offense, felony*.

misfeasance   the improper performance of public duties. Compare with *malfeasance, nonfeasance*.

missing person   an individual whose voluntary or involuntary absence from their usual and expected environs causes concern. See *National Center for Missing and Exploited Children*.

mistaken identity   a defense based on the assertion that the victim or witness has erroneously identified the *defendant* as the *offender* in a crime.

mistake of law   errors made by an individual in understanding how the law applies to their behavior.

mistrial   the termination of a *trial* early due to factors such as jurisdiction issues, improper conduct of an attorney, or a *hung jury*.

mitigating circumstance   a circumstance surrounding the commission of a crime which argues for leniency in sentencing. Compare with *aggravating circumstance, extenuating circumstance*.

M'Naughten rule   an *insanity* test named for Daniel M'Naughten who, in 1843, committed a high-profile murder in England. In his defense, M'Naughten claimed he was insane at the time of the crime and consequently should not be held responsible for his actions. He was found not guilty by reason of insanity. For years after this ruling, those accused of violent crime used the M'Naughten rule as a defense. See *insanity defense*.

M.O.   abbreviation for *method of operation or modus operandi.*

mob   a group of persons whose disorderly or violent behavior threatens public order and safety. Also a term used for *organized crime.*

mobile drug lab   a *clandestine drug lab* used to produce drugs like *methamphetamine*, which is housed in a truck, van, or other motor vehicle.

Mobilization for Youth   an initiative of President Kennedy's administration intended to address the problems underlying urban delinquency by providing special opportunities to disadvantaged youths.

mob journalism   media reports that convey popular opinions of criminal cases rather than information based on verifiable facts.

mob justice   the practice by a *mob* of meting out punishment to those they believe have committed a crime. See *lynching.*

mobster   a criminal affiliated with *organized crime.* See *Mafia.*

mock prison   a prison-like setting where students or other experimental subjects experience the organizational and interpersonal dynamics of prison life. Perhaps the most famous mock prison was that created by social psychologists at Stanford University in the early 1970s where students played roles as either guards or inmates. The experiment was controversial because those playing guards became extremely brutal and those in the inmate role suffered psychological harm.

Model Penal Code   a criminal code developed and recommended by the *American Law Institute* in 1962.

modified index   the *crime index* plus the offense of *arson.*

modus operandi (M.O.)   see *method of operation.*

mole   an *undercover* operative who collects information to use in subsequent investigations or prosecutions.

molestation   the making of improper or illegal sexual advances, most often where an adult victimizes a child. Molestation frequently involves a violation of trust.

molester   one who engages in *molestation.*

money laundering   the practice of infusing ill-gotten funds into legitimate businesses so the money cannot be easily traced by authorities. See *audit trail.*

Monitoring the Future   sponsored by the *National Institute on Drug Abuse*, a large-scale study of the behaviors, attitudes, and values of American secondary school students. Monitoring the Future includes details on drinking and illegal drug use.

M

moral panic   the elevation of a social issue that previously received little attention to a matter of national moral urgency.

mores   *norms* that carry greater moral significance in society. Compare with *folkways*.

Morris, Norval (1923–2004)   a prominent professor of law and *criminology* at the University of Chicago known for his writings on *sentencing, corrections,* and *crime control*. His published works included *The Honest Politician's Guide to Crime Control* and *The Future of Imprisonment*.

Mothers Against Drunk Driving (MADD)   a national organization dedicated to fighting drunk driving through publicity campaigns and the passage of legislation. Many members of MADD have lost loved ones due to drunk driving accidents. MADD is composed not only of mothers, but also fathers and other individuals concerned with stopping drunk driving and underage drinking. See *driving under the influence*.

motion   a request of a *judge* or *court* to rule on an issue in favor of the requestor.

motion in limine   a motion requesting that certain *evidence* not be introduced in *court*.

motivated offender   in *routine activities approach*, an individual with a desire to commit a crime.

motive   the reason a perpetrator has for committing an offense. Compare with *means* and *opportunity*.

motor vehicle theft   the theft of an automobile, truck, van, or other motor vehicle. Also referred to as *auto theft*. Due to the value of most motor vehicles, such an offense is generally a *felony*. Motor vehicle theft is an *index crime* as defined by the *Federal Bureau of Investigation (FBI)*. Compare with *carjacking, joyriding*.

moulage   a mold of shoeprints, tire tracks, or others impressions made with plaster of Paris or other material to preserve *evidence* for use in criminal cases. See *dental stone, impression evidence*.

muffler   same as *silencer*. Also referred to as a suppressor.

mugger   a street robber who often assaults his victims.

mug shot   a photograph of an *offender* usually taken during *booking*.

Muhammad, John Allen (1960–2009)   with *codefendant Lee Malvo*, he was convicted of perpetrating the *Beltway Sniper* shootings in 2002. Muhammad was executed by lethal injection in 2009. See *sniper*.

mule   a person, most often a woman, who secretly transports drugs from one location to another.

multi-level model a model of crime or justice-related variables composed of multiple levels of analysis. An example of a multi-level model is one that includes individual youths, the schools they attend, and the neighborhoods where they live.

M

multiple homicide the killing of many victims by an offender or set of offenders within a short period of time. Multiple homicide includes *mass murder*, *serial murder*, and *spree murder*. Also referred to as multiple murder.

multiple personality disorder a *personality disorder* where the patient evidences two or more distinct personalities, each often with its own background, manner of speech, and personality characteristics.

multisystemic therapy (MST) a comprehensive family-based treatment for antisocial behavior. MST addresses a wide variety of factors at the individual, family, peer, and school levels. Designated one of the *Blueprints for Healthy Youth Development* by the *Center for the Study and Prevention of Violence (CSPV)* at the University of Colorado, evaluations of MST have shown it to be effective.

multivariate analysis statistical analysis employing more than one *independent variable*. A multivariate analysis of the causes of crime might include independent variables, such as parenting practices, peer networks, and risky behavior.

mummification the drying and preservation of a corpse due to burial in a dry place or exposure to a dry climate.

municipal court a lower court serving a municipality and surrounding area that hears *misdemeanor* cases.

murder the intentional taking of a human life. Compare with *involuntary manslaughter*, *voluntary manslaughter*.

Murder Castle term given to the home of notorious *serial killer* Herman Mudgett who in the late 1800s lured more than 200 female victims to his castle-like residence in Chicago where he tortured and murdered them. Mudgett was later convicted and executed. See *serial murder*.

murder for hire a killing arranged through a *contract*. See *hit, hitman*.

Murder, Inc. a group of organized criminals who carried out *contract* killings for *organized crime* in the 1940s.

murder-suicide the committing of suicide following commission of a murder.

Murph the Surf nickname given to Jack Murphy, a criminal who with an accomplice burglarized the Museum of Natural History and stole precious gems. See *burglary*.

M

music piracy   the illegal copying of music for personal consumption or resale.

mutilation   the intentional disfigurement of a person, alive or deceased. Compare with *self-injurious behavior*.

muzzle   the open end of a firearm's barrel from which the bullet exits. Compare with *butt*.

muzzle flash   the bright flash of light created by the explosion of gunpowder when a firearm is discharged.

muzzle stamp   the outline on a shooting victim's skin of the muzzle of a gun barrel. Occurs when the gun barrel is in contact with the skin, and the body cavity immediately beneath the skin allows for the infusion of gases from the discharge of the gun. See an *abrasion collar, contact wound, entrance wound, exit wound, starring, stippling, tattooing*.

muzzle velocity   the speed at which a bullet exits the *muzzle* of a firearm.

mystery   a *crime novel* where the identity of the perpetrator often is not revealed until the end of the book. Compare with *suspense novel, thriller*. Examples of mysteries include Sue Grafton's *A is for Alibi* and Elmore Leonard's *Rum Punch*.

naive check forger   forgers who are not professional criminals. According to sociologist Edwin M. Lemert, naive check forgers are impulsive, nonviolent, and often likable. Compare with *systematic check forger*. See *forgery*.

narcissistic personality disorder   a personality disorder characterized by feelings of grandiosity and entitlement and a tendency to exploit others. People with narcissistic personality disorder demonstrate a need for attention and a lack of empathy. Compare with *malignant narcissism*.

narcotics   term used to describe drugs, such as opiates, that are used to control pain and are highly addictive. See *substance abuse*.

narrative criminology   a school of thought in criminology that argues that the narratives about crime and justice constitute sources of data to be analyzed and understood, irrespective of content. See *content analysis*.

National Academy of Sciences (NAS)   a private, nonprofit scholarly group that provides advice to the United States about matters relating to science and technology. The NAS has conducted studies and issued reports related to crime and justice issues.

National Advisory Commission on Criminal Justice Standards and Goals   a distinguished body of criminal justice experts convened during the early 1970s. The National Advisory Commission issued a series of influential reports and made recommendations on the state of a number of criminal justice functions, including law enforcement, courts, corrections, and information systems. Compare with *Wickersham Commission*.

National Archive of Criminal Justice Data (NACJD)   a repository of crime- and justice-related data sets at the University of Michigan. Part of the *Inter-University Consortium for Political and Social Research (ICPSR)*, the NACJD provides no-cost access to hundreds of criminal justice data sets. The NACJD, which receives financial support from the *Bureau of Justice Statistics (BJS)*, also makes available an on-line data analysis system for its users.

National Association of Criminal Defense Lawyers   a national organization of defense attorneys that promotes due process for accused offenders regardless of their guilt.

N

National Bullying Prevention Center an initiative of Pacer, a nonprofit organization whose mission is to help change the acceptability of *bullying* behavior among youths. The center offers a variety of online resources. See *bullying prevention, Olweus Bullying Prevention Program.*

National Center for Explosives Training and Research a federal center that offers training and subject matter expertise in identifying, handling, and disarming explosives.

National Center for Health Statistics (NCHS) under the U.S. Department of Health and Human Services, a unit within the *Centers for Disease Control and Prevention (CDC)* responsible for providing statistics that ultimately improve the health of Americans. The NCHS conducts both ongoing and periodic surveys, and serves as a repository of a vast amount of health-related data, including those related to *intentional injury* and death.

National Center for Juvenile Justice (NCJJ) the research arm of the *National Council of Juvenile and Family Court Judges (NCJFCJ)*. Located in Pittsburgh, Pennsylvania, the NCJJ is a nonprofit organization that engages in applied research, legal research, and systems research related to juvenile justice.

National Center for Missing and Exploited Children a center dedicated to locating missing children around the United States. The National Center for Missing and Exploited Children was launched in part by the highly publicized abduction and murder of *Adam Walsh*. See *Walsh, John.*

National Center for State Courts a nonprofit organization founded in 1971 that provides leadership and service to state courts. Important issues for the NCSC include assisting courts with court administration, case flow management, and court technology. Their research agenda has included topics, such as community-focused courts, *alternative dispute resolution,* and domestic relations. The National Center for State Courts is headquartered in Williamsburg, Virginia.

National Center for the Analysis of Violent Crime (NCAVC) a division of the Federal Bureau of Investigation's *Critical Incident Response Group (CIRG)*. The NCAVC investigates unusual or repetitive crimes. Its programs include *VICAP.*

National Center for Victims of Crime (NCVC) a national organization committed to helping crime victims rebuild their lives. In addition to advocating for victim rights and resources, the NCVC offers training and technical assistance.

National Center on Institutions and Alternatives (NCIA) a nonprofit organization headquartered in Alexandria, Virginia that conducts research and promotes humane alternatives for offenders. Among their services is *Client-Specific Planning.*

National Coalition to Abolish the Death Penalty a membership organization whose mission is to abolish *capital punishment* in the United States. The coalition believes that a just and humane society can exist without use of the *death penalty*. See *Bedau, Hugo Adam.*

National Commission on Correctional Health Care Standards a commission whose mission is to improve health care standards in prisons, jails, and other correctional facilities.

National Commission on the Causes and Prevention of Violence a commission convened by President Lyndon Johnson in response to the violence and civil unrest of the late 1960s. Methods for collecting relevant data included citizen surveys and the analysis of archival data.

National Consortium on Violence Research (NCOVR) an interdisciplinary and multi-institutional research, training, and resource initiative that specializes in the causes and consequences of violence. NCOVR is headquartered at Carnegie Mellon University in Pittsburgh, Pennsylvania.

National Council of Juvenile and Family Court Judges (NCJFCJ) an organization of judges dedicated to improving juvenile and family courts in the United States. The NCJFCJ conducts educational and research programs, and strives to improve the juvenile court system through a variety of programs and publications. Research for the NCJFCJ is conducted by the *National Center for Juvenile Justice (NCJJ).*

National Council on Crime and Delinquency (NCCD) a nonprofit organization that "promotes effective, humane, fair and economically sound solutions to family, community and justice problems." Established in 1907 as the National Probation and Parole Association, NCCD has taken policy positions on such issues as minorities and females in the justice system, gun violence, and the *death penalty*. In addition to advocacy work, NCCD staff conduct a number of research and evaluation studies.

National Crime Agency an agency in the United Kingdom whose purpose is to fight serious and *organized crime.*

National Crime Information Center (NCIC) a national electronic data center that contains records from domestic and foreign law enforcement and criminal justice agencies, including data on stolen autos, stolen license plates, stolen or missing guns, and wanted persons. When a law enforcement officer runs a check on NCIC, it comes back as a "hit" or a "miss." The officer thus knows if the person is wanted and dangerous.

N

National Crime Prevention Council (NCPC) a national organization that promotes the reduction of crime through addressing its underlying causes and by reducing opportunities for crime to occur. Much of the NCPC's work relates to a national anti-crime publicity campaign, including McGruff, the crime dog. The NCPC also holds several conferences each year, some focusing special attention on selected aspects of crime or delinquency prevention. The NCPC is headquartered in Washington, D.C.

National Crime Recording Standard (NCRS) a standard for maintaining the consistent recording of crime data by law enforcement agencies in the United Kingdom. Compare with *Uniform Crime Reports*.

National Crime Survey see *National Crime Victimization Survey*.

National Crime Victimization Survey (NCVS) a periodic survey of citizens conducted by the U.S. Bureau of the Census for the *Bureau of Justice Statistics*. The NCVS measures the respondents' experiences as victims of *rape, robbery, assault, burglary, larceny*, and *motor vehicle theft*. To prevent the identification of respondents, the NCVS data cannot be disaggregated. See *self-report study*.

National Criminal History Record Improvement Program (NCHIP) an initiative of the federal *Bureau of Justice Statistics* to improve the quality, timeliness, and accessibility of criminal history records.

National Criminal Justice Association (NCJA) a national nonprofit organization headquartered in Washington, D.C., whose membership is composed of *state planning agencies* and other criminal justice professionals.

National Criminal Justice Reference Service (NCJRS) a service of the U.S. Department of Justice which makes available a wide range of documents and bibliographical sources regarding crime and criminal justice.

National Deviancy Conference an annual criminology conference in the United Kingdom.

National District Attorneys Association (NDAA) a national organization that promotes the interests of prosecuting attorneys. See *prosecuting attorney*.

National Incident-Based Reporting System (NIBRS) the crime reporting program designed to replace the *Uniform Crime Reports* program of the *Federal Bureau of Investigation (FBI)*. For each criminal offense reported, law enforcement agencies participating in NIBRS collect detailed data on the offender, time, place, and other aspects of the crime. The advantage of NIBRS over summary-based crime reporting is that users can perform more detailed analyses of the correlates of crime.

National Instant Criminal Background Check System a national computerized database used to screen handgun buyers. This system grew out of the Brady Handgun Violence Prevention Act. The purpose of the instant check system is to enable law enforcement authorities to conduct background checks on those trying to purchase handguns. See *Brady Bill*.

National Institute of Corrections (NIC) a unit of the U.S. Department of Justice's Federal Bureau of Prisons that provides information, training, and technical assistance to federal, state, and local correctional agencies.

National Institute of Justice (NIJ) the research, evaluation, and development arm of the U.S. Department of Justice. NIJ awards research grants and hosts residential fellows.

National Institute on Drug Abuse (NIDA) a federal organization that conducts, supports, and disseminates research on *substance abuse* and addiction.

National Longitudinal Study of Adolescent Health also known as Add Health, this is a well-funded multiwave study of youths conducted at the University of North Carolina at Chapel Hill. The data from Add Health have been widely published in criminology, public health, sociology, and other journals. See *longitudinal study*.

National Missing and Unidentified Persons System (NamUs) a national repository of information on missing persons and unidentified deceased persons. Free DNA and other forensic services are available through NamUs. See *missing person*.

National Night Out an annual *crime prevention* event sponsored by the National Association of Town Watch. The purpose of National Night Out is to emphasize the right of citizens to enjoy their communities without the fear of criminal victimization. Many local communities participate in National Night Out, sponsoring a variety of community events.

National Opinion Research Center (NORC) a social science research organization affiliated with the University of Chicago. NORC conducts numerous national and regional surveys, some of which directly or indirectly address crime and justice-related topics. One of their major projects is the *General Social Survey (GSS)*.

National Organization for the Reform of Marijuana Laws (NORML) a national membership organization dedicated to lessening the severity of laws prohibiting the manufacture, cultivation, sale, and especially the use of marijuana. See *medical marijuana*.

National Organization for Victim Assistance (NOVA) headquartered in Washington, D.C., NOVA is a nonprofit organization committed to recognizing and implementing victims' rights and services. Among its achievements are the institution of victim compensation programs and the promotion of victim input in the criminal justice process.

National Organization of Black Law Enforcement Executives (NOBLE) an organization of Black police chiefs who are dedicated to improving law enforcement and the administration of criminal justice. Noble is headquartered in Alexandria, Virginia.

National Rifle Association (NRA) a large nonprofit organization with headquarters in Virginia whose mission is to protect the rights of gun owners and sportsmen in the United States. The NRA traditionally has contributed substantial sums of monies to the political campaigns of federal and state politicians whose views align with theirs. With respect to crime control, the NRA emphasizes the responsibility of criminals for violent crime. Consequently, it advocates harsher sentences for convicted offenders as the most sensible method for stemming gun-related crimes. See *gun control, Second Amendment.*

National Science Foundation (NSF) a United States governmental agency that funds basic research in the sciences. Among NSF's programs is the Law and Social Sciences program that supports research related to crime and justice.

National Security Agency (NSA) an agency of the United States government that collects and analyzes intelligence information in order to defend the country and defeat terrorists.

National Sheriffs Association (NSA) a national organization of county sheriffs. Headquartered in Alexandria, Virginia, the NSA promotes professionalism among sheriffs and assists members in obtaining federal and state funding. See *sheriff.*

National Tracing Center a center within the federal *Bureau of Alcohol, Tobacco, Firearms, and Explosives (ATF)* with an Internet-based system that allows participating law enforcement agencies to submit firearm traces.

National Violent Death Reporting System (NVDRS) a national program of the *Centers for Disease Control and Prevention (CDC)* to promote the collection and use of a variety of data sources on violent crime, including those from death certificates.

National Youth Gang Survey a large-scale survey of law enforcement agencies regarding their knowledge of and experience with juvenile gangs.

National Youth Survey a program of self-report studies where adolescents are interviewed about their involvement in crime and other behaviors related to a delinquent lifestyle.

nature-nurture debate   an old, but ongoing debate about whether crime is due to biological factors or environmental factors. Many criminologists now concede that both sets of factors play a role in the *etiology* of crime. See *biocriminology*.

NCAVC   abbreviation for *National Center for the Analysis of Violent Crime (NCAVC)*.

NCIC   abbreviation for *National Crime Information Center*.

NCJRS   abbreviation for *National Criminal Justice Reference Service*.

'Ndrangheta   a powerful criminal organization centered in Italy. 'Ndrangheta's criminal activities include *drug trafficking* and *extortion*. Compare with *Mafia*. See *organized crime*.

near repeat analysis   analytical method based on the notion that if a crime occurs at a specific location, the odds of another crime occurring nearby increase for a period of time. See *hot spot*.

Near Repeat Calculator   a software program that performs *near-repeat analysis* of crime data.

necrophagia   the practice of eating the flesh of dead persons. Necrophagia differs from *cannibalism* in that it involves sexual pleasure.

necrophilia   the aberrant fixation with death and dead persons. Necrophilia can manifest itself in a number of ways ranging from sexual fetishism associated with coffins and other symbols of death, to sexual relations with corpses.

neighborhood watch   an organized effort by citizens to patrol their neighborhood and report suspicious activity in order to prevent crime. Research suggests that neighborhood watch programs are associated with reductions in crime. See *crime prevention*.

neoclassical school of criminology   the school of criminological thought that introduced the notion that free will could be compromised by mental defects, rendering accused offenders less legally responsible for their behavior. The neoclassical school emerged in the early to mid-1800s with certain high profile criminal cases where the offenders claimed insanity as a defense. Compare with *classical school of criminology*.

neonaticide   the killing of a neonate or newborn child. In the late 20th century, a spate of neonaticides were committed by teenage girls who wanted to keep their pregnancies secret from their parents and others. See *child death review*.

Ness, Elliot (1903–1957)   an American law enforcement officer who gained national prominence leading a group of federal agents known as The Untouchables during *Prohibition*.

net widening   the unintended expansion of a service population. Net widening in criminal justice occurs when those persons to be diverted or otherwise served were not part of the original target population. See *collateral consequences, diversion.*

neutralization techniques   rationalizations employed by delinquents and other offenders to justify or mitigate the seriousness of their behavior. See *neutralization theory.*

neutralization theory   criminological theory advanced by sociologists Gresham Sykes and David Matza that posits that delinquents try to rationalize and minimize their acts.

neutron activation analysis   a nuclear method that permits forensic scientists to perform qualitative and quantitative analysis of solid and liquid samples of *evidence.*

new generation jail   term used to describe a *jail* that is more progressive in its design and operation. Features of new generation jails can include the elimination of barriers between staff and inmates and the development of reintegrative programming.

newsmaking criminology   the participation of criminologists in formulating crime news. An example of this is when criminologists comment in electronic and print media on high-visibility crimes. Compare with *public criminology.*

Newtown Massacre   a notorious instance of *school violence* where 20 elementary school children and six staff were shot and killed by *Adam Lanza.* See *mass murder.*

New York jack   a slang term for *heroin.*

NGRI   abbreviation for *not guilty by reason of insanity.*

NIBRS   abbreviation for *National Incident Based Reporting System.*

Nichols, Terry L. (1955–present)   one of the individuals convicted for participating in the *Oklahoma City Bombing.* Nichols is serving multiple life sentences in federal prison. See *domestic terrorism, McVeigh, Timothy.*

nightstick   a long wooden club carried by law enforcement officers. Compare with *billy club, PR-24, truncheon.*

Ninhydrin   a solution used to develop latent fingerprints on paper. See *latent prints.*

no contact order   a court order requiring an individual to have no contact with a specific person. Compare with *restraining order.*

no contest   a *plea* entered by a defendant in criminal court where there is agreement with the facts as stated in the complaint, but no admission of guilt. In most cases, a person who pleads no contest is found guilty by the court.

Noguchi, Thomas (1927–present) former chief medical examiner for Los Angeles County in California. Noguchi is known for having determined the *cause of death* of numerous celebrities, including victims Robert F. Kennedy and Sharon Tate. His books include *Coroner* and *Coroner at Large*.

noir crime novel a form of *crime fiction* where an alienated protagonist struggling with various personal demons operates in a dark world, often failing to conquer the demons or save the victim. Compare with *mystery, police procedural, suspense novel, thriller*.

no knock law a law that permits law enforcement authorities to enter a premises without notice to the occupants. Also known as the no knock rule.

no knock warrant a search warrant that permits law enforcement officers to enter the premises in question without knocking and explaining their purpose to the residents. No knock warrants are used in cases where authorities fear the residents will destroy evidence or pose a serious risk to the safety of the officers.

nolle prosequi a disposition of a criminal case where the prosecutor has decided not to proceed further. In the case of plea bargaining, it is not uncommon for prosecutors to drop one or more charges in return for a plea of guilty to the remaining charges. Nolle prosequi may also be used in cases where the prosecutor deems the available *evidence* is insufficient to obtain a conviction. See *plea negotiation*.

nolo contendere see *no contest*.

nomadic killer a *serial killer* who roams from location to location to find and kill victims. Compare with *territorial killer*. See *serial murder*.

non-capital case a serious crime not eligible for the *death penalty*. Compare with *capital offense*.

nonfeasance the failure to carry out official duties. Compare with *malfeasance, misfeasance*.

nonintervention an expression and philosophy popularized by sociologist Edwin M. Schur in his book *Radical Non-intervention*. Nonintervention rests on the assumption that by intervening, the juvenile justice system and other such systems often do more harm than good.

non-lethal weapon a weapon which can be employed by law enforcement officers against suspects without resulting in death or serious injury. There has been substantial research and development work sponsored by the

National Institute of Justice and by the U.S. Armed Forces to design non-lethal weapons.

nonpayment of child support   a criminal charge which stems from the failure by a parent or legal guardian to pay child support that was previously ordered by a court. In some states, the nonpayment of child support—also called nonsupport—is a felony and thus punishable by a prison term. Imprisonment does not relieve the negligent parent of his or her responsibilities: if they once again fail to pay support, they can once again be sentenced to prison.

non-reporting probation   a form of *probation* where the probationer does not have to check in with the *probation officer*. Offenders on non-reporting probation most often are minor offenders or others who are perceived as posing little *recidivism* risk. In such cases, probation officers have little to do other than periodically run record checks for possible violations.

norm   a standard of conduct in society. In most cases, crime violates social norms.

Northwestern University Center for Public Safety   a center located in Evanston, Illinois that offers advanced training for law enforcement, forensic, and traffic safety personnel.

not guilty   a plea expressing that the defendant did not engage in the conduct in question.

not guilty by reason of insanity   a plea and disposition which states that the defendant cannot be legally responsible for the crime charged due to not knowing right from wrong.

nothing works   a mantra that conveys the belief that correctional interventions have no rehabilitative value. Nothing works began in the 1970s with the publication of an evaluation of correctional programs called *The Effectiveness of Correctional Treatment* by Robert Martinson, Douglas Lipton, and Judith Wilks. This evaluation noted that with few exceptions, the programs had little appreciable value in changing offenders. The conclusion that nothing works in corrections was later challenged by other criminologists who cited evidence to the contrary.

not in my back yard (NIMBY)   a phrase that conveys community residents' sentiments against the placement of halfway houses or offenders in their neighborhood. Of particular interest are those people who favor progressive policies, such as the rehabilitation of offenders and the community treatment of the mentally ill, who find themselves saying NIMBY if a proposal involves their own neighborhood or community.

NOVA   abbreviation for *National Organization for Victim Assistance*.

nulla poena sine lege   a Latin phase which conveys the principle that one cannot be punished for an act that is not prohibited by law. Compare with *nullem crimen sine lege*.

nullem crimen sine lege   a Latin phrase which conveys the principle that one should not be punished for an act that was not prohibited by law prior to the act. Compare with *nulla poena sine lege*.

numbers racket   an illegal lottery game where people pay to bet on certain sets of numbers in hopes of winning a large sum of money. Often referred to as a poor man's lottery, the numbers racket generally is found in low socioeconomic neighborhoods where those with little chance of bettering their financial situation hang their hopes on winning in the numbers racket. These rackets are typically operated by *organized crime*.

Nuremberg Principle   the principle that those faced with carrying out orders should disobey those orders if they are unjust. The Nuremberg Principle stemmed from the post-World War II trials of Nazi war criminals, many of whom claimed that their crimes were the result of simply carrying out the orders of superiors. However, the magnitude and horror of the crimes committed argued not only for criminal responsibility, but also for severe penalties. See *obedience to authority*.

Nuremberg War Crime Trials   post-World War II trials of Nazi war criminals. The Nuremberg War Crime Trials resulted in convictions, life sentences, and death sentences.

obedience to authority   the phenomenon of individuals obeying an authority's orders even when those orders violate ethical or moral standards or human rights. An example of obedience to authority is when many Germans followed Nazi directives to carry out mass murders. Compare with *Nuremberg Principle.*

objection   a formal call to the court to disagree with an adversary's statement.

obscene phone call   a telephone call made by one whose purpose is to utter obscenities, sometimes for sexual gratification.

obscenity   an obscene act or written or spoken word. Compare with *pornography*. See *community standards.*

obstruction of justice   the intentional hindering of an official investigation or subsequent related justice processes.

occult crime   crime perpetrated by a person or persons who practice the occult arts. An example of an occult crime is the practice of human sacrifice by those who hold black masses, worship Satan, or who otherwise engage in the black arts. See *goths, satanism.*

occupational crime   a crime committed by persons during the performance of their legitimate occupation. An example of an occupational crime is *Medicaid fraud* by physicians. Compare with *white-collar crime.*

occupational deviance   deviant behavior engaged in during the course of a legitimate occupation. *Scientific misconduct* is an example of occupational deviance. Compare with *occupational crime.*

off-duty weapon   a firearm, most often a *handgun*, carried by a law enforcement officer when not on duty. Compare with *service weapon.*

offender   one who offends by violating the law. A criminal.

offender-based transaction statistics (OBTS)   criminal justice data that document the full range of an offender's transactions in the criminal justice system, including arrest, prosecution, trial, sentencing, and corrections. OBTS data permit policymakers to analyze such topics as the effects of proposed or enacted criminal justice legislation. The advent of computerization in criminal justice as well as the ability of disparate data systems to talk to each other has facilitated the practical use of OBTS.

offense   a crime.

offense principle   the notion that criminal law should express the moral sentiments of society. Compare with *harm principle*.

offense statistics   data on crimes reported to law enforcement authorities. Compare with *arrest statistics*.

Office for Victims of Crime (OVC)   a branch of the U.S. Department of Justice's *Office of Justice Programs (OJP)*. The OVC administers grant programs intended to assist victims and victim advocate organizations in the state and territories. See *victim, victim assistance program*.

Office of Community Oriented Policing Services (COPS)   within the *Office of Justice Programs* of the U.S. Department of Justice, the federal office charged with promoting and supporting community policing. The COPS office oversees several grants programs, including those that facilitated the hiring of President Bill Clinton's promised 100,000 community policing officers by local law enforcement agencies. It also funds a number of regional community policing institutes around the country.

Office of Global Criminal Justice   an office within the U.S. Department of State that focuses on issues such as *genocide* and *crimes against humanity*.

Office of Justice Programs (OJP)   the branch of the U.S. Department of Justice that administers federal funding programs related to crime and justice. OJP units include the *Bureau of Justice Assistance (BJA)*, the *Bureau of Justice Statistics (BJS)*, the *National Institute of Justice (NIJ)*, the *Office for Victims of Crime (OVC)*, and the *Office of Juvenile Justice and Delinquency Prevention (OJJDP)*.

Office of Juvenile Justice and Delinquency Prevention (OJJDP)   a branch of the U.S. Department of Justice's *Office of Justice Programs*. OJJDP is charged with promoting a variety of programs to reduce juvenile delinquency and to improve the administration of juvenile justice. Authorized by the Juvenile Justice and Delinquency Prevention Act of 1974, OJJDP administers discretionary and formula grant programs.

Office of National Drug Control Policy (ONDCP)   the federal office located administratively within the Executive Office of the President charged with developing and overseeing the administration's anti-drug policies. Emphases of the ONDCP are the reduction of drug use, manufacturing, and trafficking. The ONDCP has been the subject of controversy, largely because of its emphasis on drug interdiction and *supply reduction*.

Office of Sex Offender Sentencing, Monitoring, Apprehending, Registering and Tracking (SMART)   located within the U.S. Department of Justice's *Office of Justice Programs*, an office that provides information and technical assistance related to sex offending and associated issues.

officer-involved shooting   an incident where a law enforcement officer shoots someone, whether justified or unjustified. Officer-involved shootings are frequently presented to grand juries.

Ohlin, Lloyd (1918–2008)   an American sociologist and criminologist who spent the latter part of his career as a professor at Harvard Law School. With Richard Cloward, he coauthored *Delinquency and Opportunity: A Theory of Delinquent Gangs*.

Oklahoma City Bombing   the 1995 bombing of the Alfred P. Murrah federal office building in Oklahoma City, Oklahoma. The Oklahoma City bombing resulted in 168 deaths and numerous injuries. It is considered the worst instance of *domestic terrorism* on United States' soil. See *mass murder, McVeigh, Timothy, Nichols, Terry L., terrorism*.

Old Bailey   the trial court in London, England.

Old Sparky   nickname given the *electric chair*.

Olweus Bullying Prevention Program   a registered program designed to reduce *bullying* behavior and improve peer relations for students in elementary through middle school grades. There is a cost for the service. See *bullying prevention*.

omerta   the oath new members allegedly take on joining the *Mafia*. Omerta, first introduced to the public by Mafia turncoat *Joseph Valachi* in his testimony before the *McClellan Committee*, requires that the member swear never to reveal information about the Mafia or else suffer the penalty of death. A ritual involving the burning of paper allegedly accompanies the oath. See *crime family, organized crime*.

Omnibus Crime Control and Safe Streets Act of 1968   a comprehensive influential crime bill passed in the wake of the assassinations and civil disorder of the 1960s. This act provided for millions of dollars in crime control funds and established the *Law Enforcement Assistance Administration* to oversee it.

OMVI   abbreviation for operating a motor vehicle while intoxicated. Same as *driving under the influence*.

onset   the point at which a *criminal career* begins. Compare with *desistance*.

Operation Borderline   a 1988 *sting operation* of the U.S. Customs Bureau where pedophiles were arrested when they ordered *child pornography* from a phony mail-order house.

O

Operation Brilab   an FBI *sting operation* involving public corruption and labor *racketeering* in Louisiana and Texas. Those convicted included *organized crime* figures and public officials.

Operation Ceasefire   a project designed to prevent and respond to incidents of *gun violence*. Operation Ceasefire, which started in Boston, was based on *problem-oriented policing*. It was later replicated in other jurisdictions.

Operation Chaos   a special operations group of the *Central Intelligence Agency* whose purpose was to investigate foreign influence in protest activity within the United States. This assignment for the CIA from President Johnson represented a dramatic departure from that agency's foreign focus. Agents amassed files on thousands of Americans. See *political crime*.

Operation Fast and Furious   an initiative of the U.S. Department of Justice's *Bureau of Alcohol, Tobacco, Firearms, and Explosives (ATF)* where firearms were permitted to be purchased and transported to gun smugglers in order to trace the illegal trafficking of firearms. Not only did the ATF lose track of the firearms, some were used in committing crimes, including murder.

Operation Ill Wind   a *Federal Bureau of Investigation* operation focusing on Department of Defense contracts. DOD personnel colluded with defense contractors to defraud the U.S. government of millions of dollars. The investigation led to the conviction of Pentagon officials, contractors, and others. See *fraud*.

Operation Mongoose   an attempt by the *Central Intelligence Agency* to enlist organized criminals to assassinate Cuban premier Fidel Castro. See *organized crime*.

Operation Underworld   a U.S. Navy program during World War II that enlisted the assistance of *organized crime* to protect American docks.

Operation UNIRAC   an FBI *sting operation* investigating *organized crime* connected to longshoremen and shipping companies. The operation demonstrated the influence of *organized crime* in these industries.

opinion   a decision rendered by a higher court.

opioid   a type of narcotic typically used to control pain. See *substance abuse*.

Opium Wars   conflicts between the Chinese and Europeans during 1839 to 1842 over the ability to trade in opium.

opportunity theory   the notion in criminology that some offenders make a rational choice to engage in crime simply because the opportunity presents itself.

oppositional defiance disorder a disorder of children and adolescents characterized by anger, irritability, or vindictiveness. Compare with *conduct disorder*.

Oraflex a drug for arthritis marketed by the Eli Lilly Company despite knowledge that it had deadly side effects. Oraflex killed more than 100 people. See *corporate crime*.

oral argument spoken statements made in order to convince a judge or justice of a certain position.

Orchid Cellmark a private laboratory that came to prominence in the late 1980s and early 1990s as one of a few agencies and companies capable of analyzing *DNA* evidence for law enforcement. Cellmark was involved in analyzing evidence in a number of high-profile criminal cases, including *O. J. Simpson* and *JonBenet Ramsey*. See *DNA testing*.

organic brain dysfunction brain damage which can influence an individual's susceptibility to engage in criminal behavior.

organizational crime crime perpetrated with the knowledge, consent, and sometimes active participation of high-level officials within the organization. Compare with *corporate crime, white-collar crime*.

organizational deviance deviance engaged in by employees of an organization on behalf of the organization.

organized crime crime perpetrated by a formally organized syndicate or loosely organized collectivity or criminal society. See *crime family, Mafia*.

organized offender a type of *serial killer* whose crimes give evidence of a more methodical, planned offense. Organized serial killers tend to be more intelligent, better educated, and more apt to have normal relationships with members of the opposite sex than their disorganized counterparts. They often kill their victims at a location different from where the victim's bodies are found. Compare with *disorganized offender*. See *serial murder*.

Oswald, Lee Harvey (1939–1963) the man who assassinated President John F. Kennedy on November 22, 1963 in Dallas, Texas. Oswald was a former marine, expert marksman and communist sympathizer. See *assassination*.

other report study a criminological study where the researcher asks subjects about the criminal or deviant behavior of others. Compare with *self-report study, victimization survey*.

outcome evaluation the part of a *program evaluation* concerned with the results of an intervention or treatment. Compare with *process evaluation*.

outlaw a criminal, particularly one that has eluded capture.

O

outlaw motorcycle gang   an organized group of motorcyclists whose activities center on criminal enterprises. Examples of outlaw motorcycle gangs include the *Hell's Angels* and the Outlaws. See *gang*.

outsiders   a term coined by sociologist *Howard Becker* in his book by the same name to describe individuals and groups regarded as deviant by mainstream society. See *labeling perspective*.

outstanding warrant   a *warrant* that has not yet been served or executed.

overcriminalization   the excessive reach of the criminal law resulting in defining relatively minor or harmless acts as criminal. See *net widening*, *radical non-intervention*. Compare with *undercriminalization*.

overcrowding   the state where the number of jail or prison inmates exceeds the reasonable capacity of the correctional system to house or otherwise care for them. Also referred to as crowding. See *design capacity, rated capacity*.

overdose   an amount of a drug that exceeds the recommended or safe dose. An overdose of a drug can kill or cause serious illness. See *substance abuse*.

overkill   the practice of continuing to shoot, stab, or bludgeon a person after death has obviously occurred. Evidence of overkill can suggest extreme rage.

paddy wagon   see patrol wagon.

paints and polymers   a specialty of *forensic science* that focuses on the analysis of *evidence* involving paints and polymers.

Palestinian Liberation Organization (PLO)   an organization of Palestinian nationalists dedicated to the liberation of Palestine. The PLO maintains the right to use violence as a means to achieve its ends. See *terrorism*.

Palmer raids   post-World War I arrests and deportation of suspected communists. *J. Edgar Hoover* is believed to have been instrumental in the Palmer raids.

palm print   image of the human palm taken for identification purposes. Compare with *fingerprint*.

Pan Am 103 Bombing   the *terrorist* bombing and destruction of Pan Am Flight 103 over Lockerbie, Scotland in 1988 that killed 259 passengers and crew and 11 people on the ground. See *terrorism*.

panel study   a type of *longitudinal study* where the same group of individuals is administered repeated measures at different points over time in order to gauge changes in their behaviors or opinions.

panopticon   a prison whose design is credited to *Jeremy Bentham* where multitiered cells are built around a hub so that correctional personnel can view all inmates at the same time.

pansexual   one who engages in sexual relations with members of the same sex, the opposite sex, and animals.

Panzram, Carl (1991–1930)   a *serial killer* who was responsible for at least 22 murders and other crimes, including arson, burglary, and rape. Panzram was executed by *hanging*. See *serial murder*.

paper hanging   the purposeful execution of checks or similar instruments with knowledge that there are insufficient funds to cover them. Compare with *check kiting*, *passing bad checks*.

paper trail   the chain of documents that links an *offender* to the offense, especially in financial crimes like embezzlement. Compare with *evidence trail*.

paraffin test   a test used by criminalists to determine if an individual has recently discharged a firearm. When someone fires a firearm, especially a handgun, minute particles of gunpowder adhere to the skin on the hand and arm. The paraffin test picks up these particles. Paraffin tests are controversial because those who have not fired a gun can also pick up gunshot *residue*.

paralegal   see *certified paralegal*.

paramilitary   a group whose structure, training, and subculture is organized along the lines of a military unit.

paraphilia   any of several sexual deviations, including but not limited to *exhibitionism*, *pedophilia*, sexual sadism, and *necrophilia*. While some paraphilias are *victimless crimes*, others represent harmful behavior.

pardon   an official decree by a president, governor, or other high-ranking official which sets aside any remaining criminal penalties. Pardons can be controversial when officials exercise this privilege on their last day in office. Compare with *commutation*.

parens patriae   the philosophy that juvenile justice and other officials assisting youths serve in the place of parents. Parens patriae emerged in response to the harsh treatment that children once received in institutions and work settings. In the latter part of the 20th century, this philosophy was called into question with the movement toward harsher treatment of juvenile offenders.

parent abuse   the intentional battery or other physical abuse of a parent by a child. Parent abuse often takes the form of *elder abuse*.

parenticide   the killing of one's parents.

Parents of Murdered Children (POMC)   the full name, National Organization of Parents of Murdered Children, Inc., often abbreviated POMC. This national organization headquartered in Cincinnati, Ohio is dedicated to providing emotional support to those who have lost loved ones to homicide. There are many local chapters around the country, each of which attempts to help members cope with the aftermath of murder. POMC depends primarily on charitable contributions for its operating funds. It also provides training to professionals whose work is touched by homicide. Compare with *Children of Murdered Parents*.

Parker, Bonnie (1910–1934)   a notorious serial robber of the 1930s who, with lover and accomplice *Clyde Barrow*, was responsible for several murders. Parker was killed in an ambush by Frank Hamer and other lawmen.

P

parole the conditional release under supervision of a convict prior to the expiration of the sentence. Parole generally is granted by a *parole board*. Upon release, the offender periodically reports to a *parole officer* who ensures that the *parolee* abides by certain conditions. Compare with *aftercare*.

parole authority the state or federal organizational unit which oversees the administration of *parole*.

parole board a body composed of professionals and lay persons who make recommendations and decisions as to the suitability of release on *parole*. Parole boards also typically hear cases of *parole violation*.

parolee one released from *prison* on *parole*. Compare with *probationer*.

parole officer an officer charged with supervising and monitoring the whereabouts of persons on *parole*. Parole officers may also prepare *parole violation* reports, and may verify placements for *parolees* eligible for release from prison.

parole revocation hearing an administrative hearing during which officials decide whether an individual's *parole* should be revoked due to a new crime or a *technical violation*.

parole violation a violation of the conditions of release on parole, including the commission of a new offense.

parole violator one on parole who has committed either a new offense or a *technical violation*.

parricide the killing of a relative.

participant observation a research methodology where the researcher mingles with the subjects of the study. An example of a participant observation study was the study of *tearoom trade* by Laud Humphreys, an American sociologist and author.

Part I offense any of eight offenses included in the *Uniform Crime Reports* crime index. Part I offenses include murder and non-negligent homicide, rape, robbery, aggravated assault, burglary, larceny, motor vehicle theft, and arson. Also referred to as Part I crime. Compare with *Part II offense*.

Part II offense any offense reported to law enforcement not included as Part I offenses. Part II offenses include, but are not limited to assault, forgery, passing bad checks, petit theft, and driving under the influence. Compare with *Part I offense*.

passing bad checks the issuance of checks when the issuer knows there are insufficient funds to cover them.

P

patent print   visible prints made when an object comes into contact with a surface and leaves an impression on that surface.

pathologist   a physician who specializes in determining the causes of death due to disease or injury. Pathologists examine organ structure and human tissue to uncover the circumstances surrounding death. Compare with *coroner, medical examiner.*

pathology   the science of diseases. See *pathologist.*

pathways to crime   the various ways that a youth moves from unruly behaviors to overtly illegal behavior. Research has shown that there are distinct pathways to crime.

patricide   the killing of one's father. One of the most infamous patricides was the *shotgun* slaying by brothers Eric and Lyle Menendez of their affluent parents in California.

patrolman   a uniformed police officer whose responsibility is to patrol a specified geographical area. Compare with *detective.*

patrol officer   same as *patrolman.*

patrol wagon   a police van capable of holding multiple arrestees. Also referred to as *paddy wagon.*

payola   a *bribe* offered in return for a favor by a politician or other official.

PCR   abbreviation for *polymerase chain reaction.*

peace circle   see *peacemaking circle.*

peacemaking circle   an alternative justice practice where offenders, victims, and other participants sit in a circle in order to bring about understanding, restoration, and healing. See *restorative justice.*

peacemaking criminology   a school of criminological thought emphasizing nonviolence and social justice.

peace officer   a law enforcement officer charged with keeping the peace.

peculate   to embezzle. See *embezzlement.*

pederasty   a sexual relationship between an adult male and a juvenile male. Compare with *pedophilia.*

pedophile   one who engages in *pedophilia.*

pedophilia   a sexual attraction to children.

Peel, Sir Robert (1788–1850)   a British statesman who is credited with establishing police services which served as the foundation for modern policing. See *bobby.*

peeping Tom   one who engages in *voyeurism*, particularly by looking in windows of residences at night.

peer network   the group of people with whom a youth socializes and interacts, and by whom the youth is influenced.

penal colony   a separate tract of land, often on an island or other remote location, to which condemned prisoners are assigned. One of the most infamous penal colonies was *Devil's Island*, the name for a penal colony in French Guiana.

penal institution   same as *prison*.

penal philosophy   preference regarding the punishment and treatment of offenders.

penal reform   organized efforts to lessen the severity or inhumanity of criminal sanctions, especially imprisonment. See *John Howard Society*.

penal sanction   any legal consequence after criminal conviction, especially those involving confinement or control.

penal servitude   serving as an indentured servant as part of a *penal sanction*.

penalty enhancement   an add-on to a sentence as a result of a specific characteristic of the offense, such as the use of a weapon or serious injury to a victim.

penitentiary   a correctional facility designed for the long-term confinement of convicted adult felons.

Pennsylvania school of criminology   a tradition of criminological research originating at the University of Pennsylvania. The most influential figures of the Pennsylvania school of criminology were *Thorsten Sellin* and *Marvin E. Wolfgang*.

Pennsylvania system   a system of corrections characterized by solitude, hard work, and reflection for the inmates.

penology   the study of punishment and corrections.

Pentagon procurement scandal   a scandal where the U.S. Armed Forces paid exorbitant prices for goods in return for kickbacks from the contractors. See *kickback*.

pepper spray   an aerosol containing capsicum used for personal protection and subduing suspects. Pepper spray causes severe irritation of the eyes and difficulty in breathing, rendering the person sprayed temporarily helpless. Compare with *mace*. See *non-lethal weapon*.

percent bond   a *bond* that requires the deposit of a percentage of the face amount of the bond, usually 10%.

P

peremptory challenge   the right during *voir dire* to question the seating of jurors without citing a specific cause. District attorneys and defense attorneys each get a certain number of peremptory challenges. See *jury, jury selection*.

perjury   lying under oath in a court of law. Perjury is a *felony* punishable by imprisonment.

perpetrator   one who has committed or is suspected of committing a crime. Also known as a perp.

persistence   the lasting quality of *criminal careers*. See *onset, desistance*.

personal crime   a crime where the offender confronts the victim. *Robbery* and *rape* are examples of personal crimes. Compare with *property crime*.

personality disorder   a stable pattern of behavior that begins early in life and persists into adulthood. Such patterns are maladaptive and frequently result in disruptions in relationships, employment, and finances. Personality disorders are detailed in the *Diagnostic and Statistical Manual of Mental Disorders (DSM)*. Examples of personality disorders associated with crime and aggressive behavior include *antisocial personality disorder* and *narcissistic personality disorder*.

personal larceny   same as *larceny*.

person of interest   a person who may be involved in a crime or simply have pertinent information relating to the crime.

Persons In Need of Supervision (PINS)   term used to describe youths who are not delinquent and as such do not require juvenile justice processing, but who need supervision and services. See *status offense*.

perversion   an unnatural proclivity toward a certain behavior, especially of a sexual nature. *Pedophilia* is an example of a perversion.

petechial hemorrhage   pinpoint hemorrhages of the eyes indicative of death by strangulation. See *strangling*.

Peterson, Scott (1972–present)   a California man convicted of killing his wife, Laci, and their unborn daughter.

petit jury   a jury, generally of 12 persons, whose role is to listen to the facts of a case in a trial and render a *verdict*. Compare with *grand jury*.

petit theft   *theft* of goods or services of an amount less than that required to qualify for a *felony*. Petit theft is a *misdemeanor*.

petty offense   a minor crime that represents little harm to society. Compare with *misdemeanor*.

petty theft   same as *petit theft*.

Pew Charitable Trusts   a private, nonprofit organization whose purposes include improving public policy. Pew Charitable Trusts have supported work to address the problem of *mass incarceration*.

phishing   the practice of sending solicitous e-mails to unknown users in order to gain trust, information, or money. See *computer crime*.

phony accident claim   see *insurance fraud*.

photo array   a group of photographs of individuals, one or more of whom are suspects that are shown to a witness in order to make an identification. Compare with *lineup*.

phrenology   the outdated study of the human skull to determine character, mental ability, and the propensity for criminal activity. Phrenologists measured human skulls, paying particular attention to bumps, indentations, and other irregularities. See *Bertillon classification system*.

physical abuse   abuse characterized by physical aggression and often resulting in injury. Compare with *psychological abuse*.

physical assault   the use of the human body or weapons to inflict injury.

physical evidence   tangible materials that offer the potential of solving a crime and bringing those responsible to justice. See *evidence*.

physical stigmata   any physical characteristic or abnormality once considered indicative of criminality. Examples of physical stigmata include a protruding brow or eyes set closely together.

physical trace evidence   same as *physical evidence*.

pickpocket   a theft offender whose *method of operation* consists of reaching into the pockets of victims to misappropriate wallets and other goods. Pickpockets employ a variety of tactics, including bumping into their victims. Compare with *cutpurse*.

pigeon   another term for *mark*. Pigeons have the reputation of being gullible, if not stupid, for being so easily duped by their victimizers. See *confidence game*.

pigeon drop   a *confidence game* where the *mark* is convinced to deposit a sum of money in order to share in a larger pot. As with most confidence games, the pigeon drop depends in large part on the greed of the mark.

pilfering   the theft of merchandise by employees.

pillory   a wooden structure with holes for the head and hands used to punish and humiliate minor violators in colonial times. Compare with *stocks*.

pimp   a man who controls a *prostitute* and lives off her earnings. Pimps are more common with street prostitutes who, often as young *runaways*, are taken in by the pimp's promises of protection and material goods.

PINS   abbreviation for *Persons in Need of Supervision*.

piquerism   a sexual proclivity toward the cutting, stabbing, puncturing, or tearing of human flesh. Piquerism often involves intense sexual gratification from engaging in such practices, including the reactions of the victim suffering such trauma, which sometimes results in death.

PIRA   abbreviation for the *Provisional Irish Republican Army*.

piracy   the act of robbing ships at sea or aircraft during flight. See *air piracy*.

pirate   one who engages in *piracy*.

pistol   a small firearm designed to be held and fired in one hand. Also, a firearm whose chamber is integral with the barrel.

pistol-whipping   the use of a handgun to strike and injure a person.

Pistone, Joe (1939–present)   a *Federal Bureau of Investigation* agent who worked *undercover* to gather *evidence* against *organized crime*.

pizzo   a protection fee paid to the Mafia by business owners in Sicily and Italy. See *protection racket*.

place manager   in *routine activities theory*, an individual whose control or monitoring of places discourages crime.

plagiarism   passing off another's written work as one's own. Plagiarism also includes the theft of ideas, in addition to their manifestation. Although plagiarism generally is dealt with administratively, it can be the subject of criminal or civil actions. See *scientific misconduct*.

plainclothesman   same as *detective*.

planting evidence   the illegal practice of placing *evidence* in a location so as to incriminate another person.

plaster cast   see *moulage*.

plea   a defendant's formal, recorded denial or admission to a charge in criminal court.

plea bargain   same as *plea negotiation*.

plea negotiation   the process and result of an agreement between a prosecuting attorney and defense counsel to reduce the seriousness or number of charges in a criminal case in return for a guilty plea and possible leniency in sentencing. Also referred to as plea bargain. See *charge bargaining, sentence bargaining*.

P

poacher   one who engages in *poaching*.

poaching   the illegal capture or killing of game or protected animals. Poaching can be extremely lucrative, so much so that poachers have been known to use violence toward enforcement authorities and others who try to stop them. See *camouflage-collar crime, wildlife crime.*

poisoning   the intentional act of administering toxic substances to another person for the purpose of harming them. See *toxicology.*

polarized light microscopy   the use of a polarized light microscope to detect the structure and composition of materials of forensic interest.

police   officials whose responsibility it is to enforce criminal laws and ensure public safety.

police academy   a facility of law enforcement agencies where officers receive training on such practices and procedures as self-defense, using firearms, preserving a crime scene, interviewing witnesses, and testifying in court.

police athletic league (PAL)   a program of police officers working with youths in athletic and other recreational activities to reduce delinquency and develop responsible citizenship and respect for the law. Originally oriented toward athletics PALs now include activities such as computer skills and homework.

police brutality   the application of *excessive force* by police on citizens. Police brutality received national attention with the 1991 beating of motorist *Rodney King* by Los Angeles police officers.

police checkpoint   a location on a street or highway where law enforcement officers stop some or all vehicles for a specific purpose, such as searching for wanted felons. One type of police checkpoint is the *sobriety checkpoint.*

police chief   in many police departments, the highest level executive. Police chiefs generally have worked their way up through the ranks from patrol officer. Compare with *police commissioner, sheriff.*

police commissioner   the highest executive in some police departments, even where there is a *police chief.*

police community support officer   in England and Wales, a uniformed civilian member of a police force that primarily engage in *community policing.*

police corruption   the acceptance of bribes and related practices by police. See *corruption, grass-eaters, internal affairs, Knapp Commission, meat-eaters, Serpico, Frank.*

police detective   a police officer, often one who wears plainclothes instead of a uniform and who specializes in the investigation of serious crimes, such as homicide, robbery, and burglary.

police discretion   the ability of law enforcement officers to use their judgment in the exercise of their duties.

police diversion   a *diversion* program operated by a law enforcement agency.

Police Executive Research Forum (PERF)   a national organization of police executives dedicated to improving policing and advancing professionalism through research and involvement in public policy debate. PERF funding depends primarily on government and private grants and contracts. Projects that PERF undertakes include analyses of *racial profiling, community policing, domestic violence, firearms trafficking* and the evaluation of *gun buy-back programs.* PERF maintains collaborative relationships with the *International Association of Chiefs of Police*, the *Police Foundation*, and other law enforcement organizations.

Police Foundation   a nonprofit organization dedicated to improving policing through research, technical assistance, and technology. The Police Foundation was founded in 1970 and has been a leading force in promoting *community policing* through its research. It maintains its headquarters in Washington, D.C.

police impersonator   an individual who dresses in a police uniform or otherwise portrays himself as a law enforcement official in order to commit a crime.

police misconduct   legal or ethical infractions committed by police. Examples of police misconduct include the use of *excessive force,* the acceptance of bribes, and the commission of perjury. See *police corruption, Serpico, Frank.*

police murder   the intentional, unjustifiable killing of a citizen by a law enforcement officer. See *police brutality.*

police officer   a member of a police force.

police procedural   a form of crime novel which details the procedures and equipment police use as they investigate crimes. Compare with *mystery, thriller, whodunit.*

police protection   security services offered by law enforcement authorities to witnesses, dignitaries, or others whose lives may be in danger. See *Witness Security Program.*

police torture   the *torture* of individuals in custody by law enforcement officers.

policy experiment   a project designed to demonstrate the merits of a new criminal justice policy.

political crime   crimes that undermine or otherwise threaten the authority or stability of a government. Political crime can be perpetrated by either domestic or foreign offenders. See *Watergate*.

political espionage   *espionage* conducted for political purposes. The *Watergate* break-in was an example of political espionage.

political policing   policing practices by law enforcement authorities to maintain control by exerting power over others.

political prisoner   a person being held prisoner primarily for ideological reasons.

polygamy   the illegal practice of being married to more than one person at the same time.

polygraph test   same as *lie detector*.

polymerase chain reaction (PCR)   a technology for amplifying small amounts of DNA. PCR is useful in analyzing biological material in criminal cases where only very small samples of *evidence* are available. See *DNA testing*.

Ponzi scheme   a *scam* where the operator promises high returns on investments to new investors while simply placating them with deposits from new recruits. Ponzi schemes are destined to collapse because they are built on fictitious premises. See *Madoff, Bernard, white-collar crime*.

population   a large group of individuals studied by a researcher in order to draw certain conclusions. Compare with *sample*.

pornography   sexually explicit or other objectionable written material or pictures. See *community standards, Meese Commission*.

positional asphyxia   a form of *asphyxia* that results when the position of the individual prevents him or her from adequately breathing.

positive identification   the verification of the identity of a *suspect, witness*, missing person, or other individual of interest to authorities. Also called positive ID.

Positive School of Criminology   the school of criminological thought which emphasized the application of scientific methods and the identification of individual factors responsible for crime. The Positive School had its origins in Italy where *Cesare Lombroso* and *Enrico Ferri* strove to isolate physical features most closely associated with criminals. Also called Positivist School.

P

positivism   an intellectual movement beginning in the late 19th century emphasizing the collection and analysis of empirical data as well as the diagnosis and treatment of pathological conditions. See *Positive School of Criminology*.

possession   the charge for having a quantity of prohibited drugs.

post-blast residue   remnants of an explosion that can be analyzed to determine the chemical composition.

post-conviction DNA testing   the use of *DNA analysis* to determine the actual guilt or innocence of an individual who has been convicted of a crime. See *Innocence Project, wrongful conviction*.

post-conviction remedies   any of various legal means of reexamining and changing a criminal sentence, including the filing of appeals.

post-exoneration offending   the recidivism of an *offender* who was exonerated of an earlier crime.

postmodernist school of criminology   the school in criminology that asserts that crime is a social construction of those in power rather than violations of agreed-on social norms.

postmortem   relating to the period after death. Also, refers to the medical examination of a deceased after death. In addition, this examination is referred to as *autopsy*.

postmortem lividity   a discoloration of human tissue caused by the effects of gravity pulling the blood to the lowest parts of the deceased. Postmortem lividity can be used to establish whether the deceased was moved after death.

postpartum defense   a defense used by some women when charged with certain violent crimes, including the murder of their newborn child. The postpartum defense asserts that during the period after giving birth, physical, mental, and emotional changes in the mother can prompt erratic behavior that under normal circumstances would not exist.

post-release control   a form of supervision for those released from prison. Post-release control differs from *parole* in that it can be either mandatory or discretionary.

posttest   a subsequent analysis of a phenomenon under study.

Post-traumatic stress disorder (PTSD)   a serious mental disorder experienced by those who have suffered severe trauma. Symptoms of PTSD include nightmares, flashbacks, and depression. PTSD has been used as a defense by those accused of violent crime, including murder. Victims of violent crime may suffer from PTSD.

P

power-control theory an explanation of delinquency which asserts that mothers occupying traditional roles exercise more control over their daughters than their sons. Thus, this control keeps the daughters from engaging in delinquency. Mothers occupying untraditional roles do not have such control over their daughters who are therefore at greater risk of engaging in delinquency.

power elite sociologist C. Wright Mills' term for those in society who possess the political and financial power in society.

power few the relatively small number of offenders who are responsible for a disproportionate amount of crime.

PR-24 a *baton* fashioned of synthetic material with a handle attached perpendicular to the main shaft. PR-24s offer advantages over traditional nightsticks and require specialized training. Compare with *nightstick, truncheon*.

praxis in *Marxist criminology*, action to bring about meaningful change.

pre-blast explosive compounds organic peroxides and other materials used to fashion explosives.

precedent a previous court decision which serves as an authority or rule to justify or authorize another case.

precipitation hypothesis the assertion that violence as reported and portrayed by the various media actually spawn further violent behavior in society. Compare with *catharsis hypothesis*. See *media violence*.

predator an *offender* who actively stalks or otherwise pursues his victims.

predatory violence violence that is the result of purposeful premeditated action. Compare with *affective violence*. See *stalking*.

prediction scales statistics-based instruments used to predict the likelihood of future offending by individuals. See *statistical prediction*.

prediction table a device used by correctional authorities to predict the future risk of recidivism by past behavior and personal characteristics. See *clinical prediction, statistical prediction, prediction scales*.

predictive policing the deployment of police resources in time and space based on sophisticated mathematical models using crime and other data. See *hot spots*.

predisposition report a report prepared prior to a youth's disposition in *juvenile court* that summarized an investigation of risk and protective factors, family history, and treatment options. Compare with *presentence investigation report*.

P

preliminary hearing   a criminal court hearing during which *evidence* is presented by the prosecution in an effort to establish whether to proceed with *felony* charges.

premeditated murder   a *murder* that is planned with conscious forethought ahead of the crime.

premeditation   the planning of a crime in advance in such a way as to show prior intent.

prescription drug   a drug that requires a prescription issued by a licensed medical professional in order to be dispensed by a pharmacy.

presentence investigation   an investigation undertaken, generally by a *probation officer*, in order to supply background information on a defendant prior to sentencing. Presentence investigations typically include family history, employment record, and military history as well as information about the instant offense, including statements from victims. See *presentence investigation report*.

presentence investigation report (PSI)   a report prepared in advance of sentencing, often by a probation officer or investigator, that summarizes the result of a *presentence investigation*. See *Client Specific Planning*, *private presentence investigation report*, *San Francisco Project*.

presentence report   same as *presentence investigation report*.

President's Commission on Law Enforcement and the Administration of Justice   a national crime commission appointed by President Lyndon B. Johnson in the 1960s. Convened in the wake of President Kennedy's assassination as well as certain civil disturbances during this period, the President's Commission summarized or commissioned research on a number of topics, such as police behavior. This commission's recommendations were instrumental in the passage of the *Omnibus Crime Control and Safe Streets Act of 1968*.

presumptive sentencing   a sentencing scheme where the legislature has specified a standard sentence, but if there are aggravating or mitigating circumstances the sentencing judge may use *discretion* to deviate from that sentence.

presumptive test   a preliminary chemical test to determine the type of substance present. Compare with *confirmatory test*.

pretest   in research, the administration of a survey, test, or other measure before the intervention of interest is administered. Compare with *posttest*.

pretrial conference   a meeting occurring prior to a criminal trial during which the prosecutor and defense attorney discuss possible plea negotiations. See *plea negotiation*.

P

pretrial detention   confinement in a *jail* or other holding facility prior to *trial.*

pretrial discovery   see *discovery.*

pretrial diversion   an alternative to prosecution, the *diversion* of those charged with crime before a trial has taken place. Compare with *police diversion.*

pretrial hearing   a court hearing at which motions are made or plea negotiations are considered. Compare with *arraignment, preliminary hearing, trial.*

pretrial publicity   media coverage which potentially threatens the ability of the defendant to have a fair trial. See *change of venue.*

pretrial release   the practice of conditionally releasing criminal defendants prior to trial without the formal posting of bail. Those who participate in pretrial release generally must have stable residence and employment, as well as other ties to the community that suggest they are likely to appear for trial. Compare with *bail.*

pretrial services   court services that work to ensure the appearance of criminal defendants. See *bond.*

Pretrial Services Resource Center   a nonprofit organization based in Washington, D.C. that provides information on pretrial issues for criminal justice professionals and others. The Pretrial Services Resource Center works to improve the criminal justice system at the pretrial stage in order to improve safety, services, and appearance rates, as well as reduce recidivism. Its services include serving as an information clearinghouse, providing technical assistance, and promoting best practices.

prevalence   the extent to which a problem like crime is found in a population. Compare with *incidence.*

preventive detention   the holding of a defendant prior to *trial* because of a perceived risk of reoffending or fleeing.

preventive patrol   patrol by law enforcement officers for the express purpose of preventing crime by maintaining visibility rather than simply reacting to citizen calls for service.

price-fixing   an illegal collusion between two or more corporations to set the prices of merchandise or services in order to minimize competition and maximize mutual profits. See *corporate crime, white-collar crime.*

prima facie evidence   *evidence* sufficient to prove a particular fact unless specifically rebutted.

primary conflict   according to criminologist *Thorsten Sellin*, the conflict of one culture's norms with another. Compare with *secondary conflict.*

P

primary crime scene   the *crime scene* in a homicide case where the body is found. Compare with *secondary crime scene.*

primary deviance   in the *labeling perspective*, the conduct which originally caused the labeled person to come into contact with authorities. Compare with *secondary deviance.*

primary high explosive   an explosive that is sensitive to heat and shock. Primary high explosives are used to detonate a *secondary high explosive.*

primary prevention   the prevention of delinquency before the onset of behaviors characteristic of delinquency. Compare with *secondary prevention, tertiary prevention.* See *crime prevention.*

principal investigator   the member of a research team with primary responsibility for planning and carrying out the study.

prior record   at any given point in time, the history of arrests and convictions of an individual. Prior record often is used by criminal courts as one criterion to determine appropriate sentences for defendants. Compare with *rap sheet.*

prison   an institution for the long-term confinement of convicted felons. Compare with *jail, lockup, workhouse.*

prison camp   a correctional facility located in a rural area whose physical facilities consist of barracks, tents, or similar temporary construction. Prison camps may also have special purposes for inmates, such as outdoor work.

prison colony   see *penal colony.*

prison crowding   the condition where the number of inmates in a prison exceeds an established standard. The issue of prison crowding came to prominence in the United States in the 1970s and 1980s after changes in sentencing laws limited or eliminated parole, thus keeping offenders in prison longer. Also, referred to as prison overcrowding. See *design capacity, rated capacity, prison population.*

prisoner   a person deprived of liberty, especially one confined in a *prison* or *jail.*

prisoner rights   basic rights that prisoners should have, including safety and healthcare. Compare with *civil rights.*

prison farm   a working farm where prison inmates perform agricultural duties, such as cultivating crops and raising livestock.

Prison Fellowship Ministries   an organization founded by *Watergate* figure Chuck Colson. Prison Fellowship Ministries, headquartered in Merrifield, Virginia, promotes the spiritual healing of offenders, their families, the victims, and the entire community.

prison industrial complex   the extensive financial and political enterprise represented by the correctional system.

prison industry   manufacturing facilities operated by correctional systems that employ inmate labor.

prisonization   generally, the adoption by inmates of the customs, codes, and culture of a prison. Compare with *institutionalization*.

prison overcrowding   see *prison crowding*.

prison population   the total number of inmates in a prison or correctional system at any given time. See *design capacity*, *rated capacity*, *prison crowding*.

prison population forecast   the process of statistically estimating the future number of prison inmates based on demographic and crime data. Also, the results of such a process. See *ARIMA*.

prison reform   organized efforts to improve the conditions of prisons and assert the rights of prisoners.

prison riot   an often violent uprising by inmates within a prison. There are many causes of prison riots, including abuse by correctional staff and substandard living conditions. Notable prison riots in the United States include those that occurred at Attica State Prison in 1971 and New Mexico State Prison in 1980.

prison ship   in past centuries, a ship where convicted prisoners were confined.

prison suicide   the suicide of a prison inmate. Compare with *jail suicide*.

prison violence   violence that takes place within a *prison*. While prison violence generally conveys violence by inmates against guards or against one another, it can also encompass violence perpetrated by corrections officers against inmates.

private detective   a person not affiliated with a government agency who conducts investigations into criminal or civil matters for monetary gain.

private eye   another term for *private detective*.

private investigator   see *private detective*.

private presentence investigation report   a *presentence investigation report* prepared by someone outside of government on behalf of a defendant. Private presentence investigation reports became popular with the realization that government PSIs occasionally contained unverified information and did not necessarily advance the interests of the person being sentenced. See *Client Specific Planning, San Francisco Project*.

P

private security   protection services provided by a nongovernmental firm. Private security provides a substantial amount of public safety services in the United States. See *privatization.*

privatization   the assumption by private companies of criminal justice roles or functions traditionally the province of government. Privatization in criminal justice has occurred extensively in policing and corrections. See *Corrections Corporation of America.*

proactive aggression   aggression without provocation. Compare with *reactive aggression.*

proactive policing   a style of policing where officers do not simply react to calls for service, but instead seek opportunities to prevent crime, apprehend criminals, and otherwise serve the community. Compare with *reactive policing.*

probable cause   a standard which requires that a reasonable individual has sufficient reason to believe that a person committed a crime.

probation   the suspension of a sentence of a convicted offender and granting of freedom for a period of time under specified conditions. Probation, which generally is granted in lieu of confinement, is often confused with *parole.* See *Augustus, John, probation officer.*

probation contract   an agreement between a probation officer and a probationer specifying that conditions have to be met in exchange for certain considerations.

probationer   a convicted offender on *probation.* Probationers must abide by certain conditions imposed by the sentencing judge.

probation officer   an officer of a court or state agency who conducts *presentence investigation reports* and supervises convicted offenders on *probation.* See *presentence investigation.*

probation violation   an infringement of the rules of probation which can result in the revocation of probation and the reinstatement of the suspended sentence. A probation violation can stem from a new criminal charge or from a *technical violation.*

probation violator   one who has committed or is suspected of having committed a *probation violation.* Probation violators may be brought before the court, which can result in revocation of probation and reinstatement of the suspended sentence.

problem-oriented policing   a philosophy and practice of policing that emphasizes the systematic analysis of neighborhood problems in order to better allocate resources and respond to the community. Problem-oriented policing is most closely associated with law professor Herman Goldstein, author of *Problem-Oriented Policing.*

problem-solving courts   see *specialty court.*

pro bono   performed for the good of the public. Often used to describe donated legal services, including those provided for indigent defendants.

procedural justice   the fairness of a process to achieve an outcome. Compare with *distributive justice.*

process evaluation   a form of evaluation designed to assess the way a program is carried out as opposed to focusing on its results. Compare with *outcome evaluation.*

professional crime   crime committed by people who make their living by committing theft. Professional crime is distinguished from occasional crime in the identity of the offender with the criminal enterprise as their occupation. Compare with *white-collar crime.*

professional thief   a person who makes his or her living stealing. The criminologist *Edwin H. Sutherland* conducted an in-depth study of this type of offender which he wrote about in his book *The Professional Thief.*

profiler   one who performs *profiling.* See *Brussel, James, criminal investigative analysis, Ressler, Robert.*

profiling   the practice of using past empirical data on offenders and their behavior to identify, track, or apprehend other such offenders. The *Behavioral Analysis Unit* of the *Federal Bureau of Investigation* has refined the practice of profiling in serial homicide, arson, sexual assault, and other such crimes. Profiling, however, has come under fire in recent years. Many offenders who are stopped by law enforcement authorities fit a racially or culturally based profile. Recent cases involve Black citizens or Arab tourists who are mistaken for suspects. See *criminal investigative analysis, racial profiling.*

program evaluation   the formal collection and analysis of data in order to assess the effectiveness of a criminal justice program. See *outcome evaluation, process evaluation.*

Program on Human Development in Chicago Neighborhoods (PHDCN)   a well-funded, large-scale *longitudinal study* designed to assess the many causes of delinquency, crime, substance abuse, and violence. The PHDCN research includes individual, family, and community measures. The Harvard School of Public Health directs the PHDCN. See *collective efficacy, cumulative disadvantage, social capital.*

Prohibition   the period in the United States from 1919 to 1932 when the manufacture, distribution, sale, and consumption of alcoholic beverages were prohibited by the *Volstead Act*. Prohibition created a *black market* for alcohol, and as a result organized crime figures like *Al Capone* met the demand with illegally produced alcoholic beverages, spawning a wave of related corruption and violence. See *bootlegger, Ness, Elliot, revenuer*.

PROMIS   abbreviation for *Prosecutor's Management Information System (PROMIS)*.

promiscuity   engaging in sexual relations indiscriminately with multiple partners. Promiscuity as well as ignorance and unprotected sex were in part responsible for the spread of the AIDS epidemic of the 1980s.

propensity score matching   a statistical technique that attempts to ensure units of analysis for an observational study are comparable to those of a randomized experiment. See *program evaluation*.

property crime   a criminal offense which involves the theft or destruction of property. Examples of property crime include *motor vehicle theft* and *arson*.

property room   the room maintained by a law enforcement agency to hold property seized as a result of criminal investigations.

pro se   a Latin term that has come to mean self-representation in court.

prosecuting attorney   an attorney who pursues criminal cases on behalf of a governmental subdivision, such as a county or city.

prosecution   the initiation and pursuit of criminal charges.

prosecution witness   a *witness* whose *testimony* is sought to support the legal case made by the *prosecutor*.

prosecutor   see *prosecuting attorney*.

prosecutorial discretion   the ability of a *prosecutor* to make decisions about whom to formally charge, what will be the charges, and whether charges will be reduced or dropped.

prosecutorial immunity   the protection of prosecutors from civil liability for decisions they make in the course of carrying out their duties.

prosecutorial misconduct   inappropriate conduct by a *prosecuting attorney* that violates court rules or ethical standards of law practice, such as withholding exculpatory *evidence*.

Prosecutor's Management Information System (PROMIS)   an early management information system developed for the scheduling of federal criminal cases, the management of witnesses, and the evaluation of prosecution services. PROMIS served as a model for many of the current information systems in criminal justice.

prostitute   a person who engages in *prostitution*.

prostitution   the practice of engaging in sexual relations for money.

protection extortion   see *protection racket*.

protection order   a judicial order intended to keep a threatened or vulnerable individual safe from another individual.

protection racket   an illegal enterprise in which business owners are encouraged to pay for protection against robbery, assault, arson, or other crimes which could hurt their business. Those selling this type of protection send the message that such crimes will occur if the payments are not made. See *pizzo*.

protective factor   any positive attribute of a person at risk of offending that helps insulate against *criminogenic factors*. Compare with *risk factor*.

provocation defense   a defense in a criminal case based on the assertion that the defendant was provoked to commit the act for which he or she is being charged. Compare with *self-defense*. See *victim-precipitated crime*.

prowl car   a police cruiser that patrols neighborhoods looking for crime.

proximate cause theory   the notion in law that individuals are culpable if there is an identifiable connection between their actions and the criminal harm in question. See *felony murder doctrine*.

prurient interest   an unnatural obsession with sexual matters.

psychological abuse   nonphysical abuse that results in psychological or emotional trauma. Compare with *physical abuse*.

psychological autopsy   the art and science of analyzing the personality and mental condition of a deceased person as well as the circumstances surrounding his or her death.

psychopath   a now dated term for a person who evidences *antisocial personality disorder*. Compare with *sociopath*.

psychopathic killer   one who kills, sometimes repeatedly, with little apparent motive or remorse.

psychopathic personality   see *psychopath, antisocial personality disorder*.

psychopathology   the science and study of mental disorders.

psychopathy   a condition characterized by few inhibitions or feelings of guilt and a tendency to engage in antisocial behavior. Compare with *antisocial personality disorder*.

P

Psychopathy Check List—Revised (PCL-R)   a psychometric instrument developed by psychologist Robert D. Hare to assess the extent to which an individual possesses psychopathic tendencies and traits.

psychophysical scaling   a set of methods used to determine the magnitude of human perceptions of various stimuli. Psychophysical scaling has been used in criminology to determine the perceived seriousness of crimes and the severity of penalties.

psychosurgery   brain surgery employed to treat mental disorders. Psychosurgery includes, but is not limited to *lobotomies* and lobectomies.

PTL scandal   a *scam* perpetrated by evangelists Jim and Tammy Bakker of the PTL organization, named for their stock phrase "Praise the Lord." Using television and live appearances, the Bakkers solicited millions of dollars in donations from their loyal followers. Instead of using the donations for religious purposes, the Bakkers diverted them for their own appetites. Both were convicted on federal tax charges.

public criminology   criminology as consumed by the public. Public criminology is concerned with the ways that criminologists can inform public debates and help solve social ills. Compare with *translational criminology*.

public defender   an attorney employed by federal, state, or local government whose role is to represent indigent defendants accused of crimes. Compare with *appointed counsel*.

public enemy   a dangerous notorious criminal. At one time, the *Federal Bureau of Investigation (FBI)* used this term to describe a notorious fugitive as its Public Enemy #1.

public housing   government subsidized residences for low-income families. Public housing projects often are plagued by violence, drug trafficking, and gang activity as well as other forms of crime and social disorder. See *Cabrini Green*.

public indecency   see *indecent exposure*.

public intoxication   drunkenness in public. See *public order crime*.

public opinion   the feelings and preferences of citizens, often gauged by telephone surveys. Survey organizations like the *National Opinion Research Center (NORC)* routinely conduct surveys of public opinion, including opinions about crime and justice-related issues.

public order crime   minor offenses that cause more nuisance than harm to society. Examples of public order crimes include *disorderly conduct, public intoxication,* and *vagrancy*.

public order offense   see *public order crime.*

public procurator   a public officer that both investigates and prosecutes crimes.

pugilistic attitude   a commonly found position of those burned to death that resembles a boxer with arms extended as if engaged in a boxing match. Pugilistic attitude results from the contraction of muscles due to exposure to intense heat.

Pulse Check   a periodic data collection effort of the *Office of National Drug Control Policy (ONDCP)* to keep abreast of current trends in drug use in the United States. Pulse Check employs telephone interviews of experts who have knowledge of *substance abuse.*

punishment   a penalty meted out to a convicted offender.

punitive populism   the belief that public support for retributive penal policy guides government policy.

purge fluid   a product of decomposition which may drain from the mouth or nose of a deceased person.

pursuit   efforts made by police, most often in motor vehicles, to apprehend a traffic offender or suspect in a crime. See *high-speed pursuit.*

Purvis, Melvin (1903–1960)   an FBI agent in the 1930s and 1940s whose professional accomplishments included the apprehension of *John Dillinger.* Purvis eventually took his own life.

push-button criminology   an ideal situation where *criminology* can assist policymakers by identifying specific policy changes to prevent and control crime. Compare with *public criminology, translational criminology.*

putrefaction   the decomposition of a corpse. See *adipocere, body farm.*

pyramid scheme   an illegal moneymaking scheme where converts purportedly make money by enlisting others who pay them residual fees to join.

pyromania   the pathological desire or compulsion to set things on fire. See *fire setting, serial arson.*

pyromaniac   one who suffers from, or engages in *pyromania.* Also *fire setter.*

Quantico, VA the site of the FBI's National Academy. The FBI maintains a training facility there. It is also the site of several important sections within the FBI, including a crime laboratory, the *Behavioral Analysis Unit,* and their Critical Response Team. See *Federal Bureau of Investigation (FBI).*

quasi-experimental design an evaluative design where experimental and control groups are not randomly selected or not identical. Quasi-experimental designs are employed in criminological research because experiments using true random assignment are sometimes difficult to arrange in criminal justice settings. See *Academy of Experimental Criminology*, *experimental criminology.*

queer criminology criminology that focuses on the lesbian, gay, bisexual, transgendered, and queer (LBGTQ) community and its involvement with crime and the criminal justice system.

questioned documents documents whose authenticity is suspect.

questioned documents examination the in-depth analysis of documents with respect to their origin and authenticity.

questioned documents examiner one who performs *questioned documents examination.*

Quetelet, Adolphe (1796–1874) a 19th-century Belgian statistician who was one of the first to analyze crime statistics to discern regularities and patterns.

qui tam suit an action taken by an individual against a contractor of the U.S. government alleging fraud or other criminal conduct. Individuals who win qui tam suits are entitled to a portion of the damages. Qui tam suits have been filed in cases of *scientific misconduct.*

Racial Democracy, Crime and Justice Network (RDCJN) a network of social scientists interested in the intersection of race, crime, and justice. The RDCJN also seeks to increase participation of minorities in this area of research.

racial justice ensuring rights to racial groups that historically have been subjected to discrimination.

racial profiling the practice by law enforcement officials of stopping minorities when no offense has been committed or no reasonable suspicion exists. Racial profiling has garnered much attention, especially with the admission by prominent Black celebrities that they had been the object of such police behaviors. See *driving while Black*.

racist one who believes in the superiority of a particular race and hates members of specific racial, ethnic, or religious groups. See *Aryan Nations, hate crime, Ku Klux Klan*.

Racketeer Influenced Corrupt Organization (RICO) a federal statute which allows law enforcement authorities to charge or civilly sue criminal enterprises. RICO contains several provisions that facilitate the investigation and prosecution of organized criminal activity. See *organized crime*.

racketeering obtaining money through a variety of illegal enterprises, including but not limited to *hijacking, prostitution, drug trafficking*, and *loan sharking*.

Rader, Dennis (1945–present) a *serial killer* who was responsible for the murders attributed to the *BTK Killer*. Rader was sentence to 10 consecutive life sentences. See *serial murder*.

radical criminology a school of theoretical criminology which emphasizes that laws are made by those with power in society and that such laws are used to control certain groups. Compare with *Marxist criminology*.

radical nonintervention the notion put forth by sociologist Edwin M. Schur that juveniles would be much better off if authorities did not interfere with their lives and label them. See *nonintervention*.

Radzinowicz, Sir Leon (1906–1999) a Polish-born criminologist who was the founding director of the Institute of Criminology at the University of Cambridge. Radzinowicz worked with other prominent criminologists, including *Marvin E. Wolfgang.*

Raman spectroscopy a type of spectroscopy that makes use of fundamental modes of vibration in the forensic analysis of *trace evidence.*

Ramsey, JonBenet (1990–1996) a 6-year-old girl and regular beauty pageant contestant from a prominent Denver, Colorado family who was murdered in her own home in December 1997. Despite a lengthy investigation, including the consultation of nationally known forensic and behavioral experts, the crime remains unsolved.

RAND Corporation a private research and development corporation based in Santa Monica, California that conducts criminal justice research and evaluation. RAND has maintained an active criminal justice research program since the early 1970s. RAND's research contributions include studies on *probation,* and programs on *crime prevention* and *three strikes and you're out.*

random breath testing alcohol breath testing at an unannounced time or place. Random breath testing is employed by law enforcement to prevent *drunk driving.*

randomization in social experiments, the allocation of subjects between experimental and control groups such that each has an equal probability of assignment to either group. See *Academy of Experimental Criminology, control group, experimental group.*

ransom money demanded by kidnappers and sometimes paid by the victim's family for their safe release. See *kidnapping.*

ransom note a note left by kidnappers spelling out their demands and the consequences of not meeting them. See *kidnapping, ransom.*

rape sexual intercourse by force, threat of force, or deception. Compare with *statutory rape.*

rape kit a kit containing products to facilitate the collection and preservation of semen, hairs, and other physical evidence taken from a *rape* victim. Rape kits are typically used by doctors and nurses working in emergency rooms where rape victims are taken for treatment. See *sexual assault forensic examination (SAFE), Sexual Assault Nurse Examiner (SANE).*

rap sheet an informal term for an offender's prior arrests and convictions.

rated capacity the capacity of a prison, jail, or other correctional facility based on standardized criteria. Compare with *design capacity.* See *overcrowding.*

rational choice model   see rational choice theory.

rational choice theory   a theory in criminology that rests on the assumption that individuals rationally weigh the pros and cons when deciding whether to engage in crimes which meet their basic needs.

Ray, James Earl (1928–1998)   the *assassin* convicted of shooting and killing Rev. Dr. Martin Luther King in Memphis, Tennessee in 1968. See *assassination*.

razor ribbon   a coiled wire bearing pointed, razor-sharp barbs used on the top of walls and fences of prisons and other correctional facilities to prevent escape.

reaction formation   a process described by criminologist Albert Cohen where young delinquents who are excluded from mainstream society engage in antisocial behavior to gain status in the eyes of their peers.

reactive aggression   *aggression* expressed in response to a stimulus. Compare with *proactive aggression*.

reactive policing   a style of policing where officers respond to calls for service. Compare with *proactive policing*.

reasonable doubt   doubt that a reasonable person would have as to the alleged guilt of an accused person being tried for a crime. A jury must find the defendant guilty beyond reasonable doubt in order to convict. This standard is higher than that employed in civil cases.

reasonable suspicion   when deciding whether a crime has been committed, a standard centered on what a reasonable person would conclude based on available facts. Reasonable suspicion is less stringent than *probable cause*.

receiving stolen goods   same as *receiving stolen property*.

receiving stolen property   the receipt of property known or likely to be stolen. Also, the charge for engaging in such an activity.

recidivism   engaging in unlawful behavior by those who previously were sentenced.

recidivist   one who relapses into criminal behavior.

Reckless, Walter C. (1899–1988)   an American criminologist best known for developing *containment theory*. Educated at the University of Chicago, Reckless spent his academic career at Vanderbilt University and The Ohio State University. A product of the *Chicago School of Criminology*, Reckless worked with *Simon Dinitz*.

reconciliation   the process where offenders and victims reach a mutual understanding about the harm caused by an offense and how it should be addressed. See *victim-offender reconciliation program (VORP)*.

R

record   the history of an individual's official involvement with the criminal justice system. Informally called *rap sheet*.

record piracy   the illegal reproduction and distribution of copyrighted sound recordings. Also called bootlegging. The material may be reproduced in any of several formats, including vinyl, cassette tape, and digital compact discs. The advent of digital sound technology and computerized desktop publishing has made it possible for record pirates to manufacture albums whose sound recordings, art work and packaging are virtually indistinguishable from the legitimate product.

recreational killer   a *serial killer* whose primary motivation to kill is for pleasure.

recreational law enforcement   policing activities in areas where the public engages in recreational activities, such as hiking, snowmobiling, boating, or skiing.

red-handed   in the act of committing a crime.

redirect examination   the examination in court of a witness by the party whose case required the witness's *testimony*. Compare with *cross examination*, *direct examination*.

red light district   a section of a city where there is a large number of houses of *prostitution*.

reentry   the postrelease adjustment of a prison inmate. Also, the movement in criminal justice concerned with such adjustment.

reentry court   a *specialty court* designed to facilitate the successful return of an offender from prison. Reentry courts can sentence an offender to a prison term with the provision that they eventually are released under the supervision of the court. One advantage of reentry courts is the early identification of the offender's risks, needs, and assets so they can be addressed prior to sentencing, during confinement, and on release from prison.

referee   a legally trained court official who hears minor cases. Compare with *judge*, *magistrate*.

reformatory   a state correctional facility intended to rehabilitate youthful or otherwise non-serious offenders. Reformatories made their first appearance in the 1800s. Compare with *prison*, *penitentiary*.

regicide   the killing of a king or queen. See *assassination*.

regional crime lab   a crime laboratory that serves a specific region of a state or territory.

Regional Information Sharing System (RISS) a program consisting of six regional centers that share criminal intelligence information. Numerous federal agencies participate in RISS, including the *Federal Bureau of Investigation (FBI)*, the *Drug Enforcement Agency (DEA)*, the *Internal Revenue Service (IRS)*, and the *Bureau of Alcohol, Tobacco, Firearms and Explosives (BATF)*, as well as thousands of other law enforcement and public safety agencies.

rehabilitation a rationale for punishment which emphasizes correcting *offender* behavior through treatment. Compare with *deterrence, incapacitation, retribution*.

rehabilitation order a community sentence for youths which gives the court flexibility in mandating necessary services.

reintegration a rationale for punishment that emphasizes returning the offender to the community in order to facilitate linkages with family and community. Compare with *rehabilitation*.

reintegrative shaming public shaming that has positive effects for the *offender*. Compare with *disintegrative shaming*. See *shaming penalties*.

Reiss, Jr., Albert J. (1922–2006) a prominent American sociologist and criminologist. Educated at the University of Chicago, he spent most of his career at the University of Michigan and at Yale University. Among his contributions were in-depth studies of police behavior.

relapse prevention treatment services designed to prevent offenders from once again engaging in the maladaptive behavior that led to their legal problems. Relapse prevention is an especially important part of correctional programming for difficult-to-treat offenders, such as sex offenders and drug addicts.

release on recognizance (ROR) the release of a charged offender without the posting of a monetary bond or surety. With ROR, the offender promises to show up at future court appearances. See *pretrial release*.

remand prison a *prison* for those who have not yet been sentenced. Compare with *jail*.

remand prisoner a prisoner confined in a *remand prison*.

reparation the act of making amends for a crime. Reparation can take many forms, including *restitution* and *community service*.

repeat offender see *career criminal*.

replica gun a nonfunctioning firearm whose appearance is similar to that of a working firearm.

R

reported crime   crimes reported and therefore known to the police. For participating law enforcement agencies, reported crimes are included in the *Uniform Crime Reports (UCR)* or *National Incident Based Reporting System (NIBRS)*. Compare with *hidden crime*.

reprieve   the delay of a punishment.

residency restriction   a law or administrative rule that prohibits convicted offenders from residing in or near certain locations, such as schools or parks. The practical application of residency restrictions, which are common for convicted sex offenders, means there are few places they can live.

residential burglary   the burglary of a private home. Compare with *commercial burglary*.

resisting arrest   the refusal to comply with a law enforcement officer's attempt to detain. Also the formal charge for such conduct. Resisting arrest as a charge usually accompanies one or more other charges that prompted the arrest like *disorderly conduct*.

Ressler, Robert (1937–2013)   an FBI agent who was an early pioneer in offender *profiling*. With fellow FBI agent John Douglas and nursing professor Ann W. Burgess, Ressler co-authored the *Crime Classification Manual*. See *criminal investigative analysis*.

restitution   money paid by an offender to victims to compensate them for injuries or loss of property. Restitution is generally combined with other legal consequences, such as *probation*, fines, and the payment of court costs.

restorative justice   a non-retributive justice philosophy which emphasizes understanding and healing for the victim, the offender, and community rather than *retribution*.

restraining order   a court order requiring one individual to stay away from another. Restraining orders are commonly used by courts in cases of *domestic violence* to keep an abusive husband away from his wife. Compare with *civil protection order*.

restriction fragment length polymorphism (RFLP)   a forensic technique that permits the analysis of *DNA* evidence from crime scenes. See *DNA testing*.

retail theft   theft taking place in retail stores.

retaliation   getting even for a real or perceived wrong or injustice. Compare with *retribution*.

retaliatory violence   violence motivated by the perpetrator's desire to get even with the victim.

retreatism one of several modes of adaptation proposed by sociologist *Robert K. Merton* in his theory of *anomie*. Retreatism occurs when individuals reject societal goals and withdraw from society either physically or through *substance abuse*.

retribution the rationale for punishment which expresses society's need for *revenge*. Compare with *deterrence, incapacitation, rehabilitation*.

revenge the motive or act of getting even for some real or perceived wrong. Revenge expresses itself in the criminal law as *retribution*. See *vigilantism*.

revenuer a colloquial term used by bootleggers to describe an agent of the *Internal Revenue Service* that investigated violations of liquor laws. See *bootlegger*.

revocation the act or an instance of something being revoked, such as *probation* or *parole*.

revolver a handgun where the bullets are contained in a rotating cylinder. Compare with *semiautomatic weapon*.

RICO abbreviation for *Racketeer Influenced Corrupt Organizations*.

ridge an elevated portion of a fingerprint. Compare with *valley*.

Ridgway, Gary Leon (1949–present) a *serial killer* who was known as the *Green River killer* in the State of Washington. Ridgway eluded law enforcement for several decades before being caught and convicted. See *serial murder*.

rifle a long gun whose barrel contains rifling.

rifling the spiral grooves inside the barrels of firearms for the purpose of causing the bullet to spin and therefore stay on a straighter course. Rifling leaves marks on the bullet that *ballistics* experts can use to match a bullet and the firearm from which it was fired. See *rifle, grooves, lands*.

Right on Crime a conservative criminal justice reform initiative in the United States which focuses on overcriminalization, juvenile justice, substance abuse, adult probation, parole and reentry, law enforcement, prisons, and victims.

right to bear arms an American right, grounded in the *Second Amendment* of the Constitution of the United States, allowing every citizen to own and possess firearms. Despite the arguments that the right to bear arms is constitutionally guaranteed, some critics counter that the need for an armed militia in the 1700s is obviated by armed forces. See *gun control, National Rifle Association*.

right to counsel the right to have legal representation before a criminal court. Also, an area of study and policy focusing on this right. In the late

20th century, the *Office of Juvenile Justice and Delinquency Prevention (OJJDP)* made the right to counsel a priority area for study and funding.

rigor mortis temporary muscular rigidity in a deceased person. Because rigor mortis is only temporary, it is used to estimate the time of death.

riot a civil disturbance, often resulting in violence. See *civil disobedience, public order crime.*

riot gun a *shotgun* with a short barrel designed to intimidate or quell rioters. Riot guns, once considered unnecessarily offensive and consequently removed from police patrol cars, have been restored in many jurisdictions due to the heavy armament of some contemporary criminals.

risk assessment the systematic process of determining the risk that individuals pose of reoffending.

risk assessment instrument a pen and paper list of items designed to collect information about an offender for the purpose of determining their future risk of reoffending. See *risk assessment.*

risk factor a factor with the potential of increasing the likelihood an individual becomes involved in delinquency. Compare with *protective factor.*

risk-focused approach the development of prevention and intervention programs for youths based on their risk of engaging in problem behaviors. Compare with *asset-focused approach.*

risk terrain modeling the analysis of local risk factors, law enforcement tactics, and crime trends to determine the most appropriate law enforcement strategy for a particular area.

ritualism in Robert K. Merton's theory of *anomie*, one of five modes of adaptation an individual may adopt in the pursuit of social goals. Ritualism consists of going through the motions with real commitment to goals.

roadblock a barrier of vehicles created by law enforcement officers to locate or stop fleeing offenders. See *Stop Stick.*

road rage extreme anger and associated violence by motorists.

robber baron a now dated term for a wealthy magnate who used power and influence to control industries through such tactics as manipulation of markets and intimidation of competitors.

robbery the act of taking money or other property from another by force or threat of force. Robbery is often confused with *burglary*. See *armed robbery.*

Rohypnol a drug used to sedate unsuspecting victims for the purpose of *sexual assault*. See *date rape.*

Rolling, Danny (1954–2006) a drifter who was responsible for the five murders attributed to the *Gainesville Ripper*, as well as other homicides. Rolling was executed by *lethal injection*.

Rome Statute the international treaty that in 1998 established the *International Criminal Court*. The Rome Statute also spelled out rules of procedure for the court.

rookie a new inexperienced recruit, generally during the first year of service. Also called newbie.

Rosenberg, Julius (1918–1953) and Ethel (1915–1953) a husband and wife who were convicted of selling U.S. atomic secrets to the Soviet Union. Both died in the *electric chair*. See *espionage*, *Walker spy ring*.

routine activities approach a theoretical perspective in criminology that suggests crime occurs when three elements converge: 1) a suitable target; 2) a motivated offender, and; 3) the absence of a capable guardian. The elimination of any one of the three elements theoretically makes the commission of a crime impossible.

Royal Canadian Mounted Police (RCMP) the national police service of Canada. Administratively, the RCMP is an agency of the Minister of Canada. In addition to serving as federal Canadian police, the RCMP also provides contract services to territories, provinces, and municipalities.

rules of evidence administrative regulations that govern the definition and use of *evidence* in criminal cases.

runaway a youth who runs away from home. Also, the charge officials level against one who runs away. Many runaways leave home to escape abuse. See *status offender*.

rural crime crime occurring in a rural area, particularly those offenses characteristic of such areas. Examples of rural crime include theft of gas from farm tanks.

rural criminology the study of crime in rural and agricultural areas. See *rural crime*.

Russian Mafia *organized crime* based in Russia or dominated by Russians. The Russian Mafia became more powerful with the collapse of communism in the Soviet Union.

rust belt any city or region where the loss of local industry has harmed the economy, raised unemployment, and brought attendant problems like crime. When factories that previously employed large numbers of people close down, many employees are put out of work and can no longer

R

contribute to the local economy. Business owners who previously depended on this commerce go out of business, leading to neighborhood decay and crime.

sabotage   intentional damage to plans or physical facilities, usually by an insider.

Sacco, Nicola (1891–1927) and Vanzetti, Bartolomeo (1888–1927)   anarchists who were convicted of murder in 1921 and executed in the *electric chair,* despite doubts about their guilt.

sadism   the derivation of pleasure from inflicting pain or suffering on others, named for the Marquis de Sade. Compare with *masochism, sadomasochism.* See *sexual sadist.*

sadomasochism   practices featuring both *sadism* and *masochism.*

safecracking   the act or process of breaking into a safe to steal the contents.

Safe Futures Initiative   an initiative designed to create a continuum of care in communities in order to prevent and control delinquency. Safe Futures is predicated on using community strengths, including the youths themselves.

safe house   a dwelling used by law enforcement or other criminal justice authorities to hide and protect witnesses or others who may be in danger. See *Witness Security Program.*

safe room   a secret room designed to protect the structure's inhabitants against attack.

safe storage laws   laws intended to require the safe storage of firearms. Safe storage laws are designed to prevent accidental injury and death due to firearms misuse.

Safety Town   a *crime prevention* program where youths learn how to be safe in their communities. Topics covered in a typical Safety Town lesson include how to cross the street, what to do in case of a fire, and what to do if approached by a stranger. Safety Towns emphasize all facets of a child's life, including home, school, and other locations and often end with graduation ceremonies for participating youths.

Saint Valentine's Day massacre   the murder of seven members of George "Bugs" Moran's gang by killers allegedly hired by mobster *Al Capone.* The murders were committed by men posing as police officers and led to increased efforts to control *organized crime* in Chicago.

same sex abuse   the physical, mental, or emotional abuse by a person toward another of the same gender.

sample   a group of cases considered representative of a *population*. Criminologists collect data on samples of offenders or ordinary citizens in order to draw inferences about the population of interest.

sanction   a legal consequence for committing a crime. Sanctions can include imprisonment, *probation*, fines, and other options.

Sandy Hook tragedy   see *Newtown Massacre*.

San Francisco Project   a series of research projects undertaken in the late 1960s and early 1970s by the School of Criminology at the University of California at Berkeley to examine practices related to *probation*. Among the studies in the San Francisco Project were inquiries into the *presentence investigation report*.

San Quentin   San Quentin State Prison, San Quentin, California. The oldest of California's prisons, San Quentin was established in 1852. It houses the state's *death row* for condemned men as well as the state's *gas chamber*, even though the condemned now face *lethal injection*.

sap   a heavy leather-clad weapon used to strike a person. Compare with *blackjack, truncheon*.

SARA model   stands for scanning, analysis, response, and assessment. A method suggested by law professor Herman Goldstein to systematically document and evaluate problems in a community in order to address them. Scanning is the collection of data on problems in a community. Police make sense of these data through analysis which leads them to a response to address identified problems. In the assessment phase, police officials evaluate their response in order to make any necessary adjustments. See *community policing, problem-oriented policing*.

satanic cult   a quasi-religious group whose practices and ceremonies center around the worship of Satan. Compare with *goths*.

satanism   the worship of Satan. See *satanic cult*.

satanist   one who worships Satan or otherwise practices the black arts.

saturation patrol   a law enforcement strategy where large numbers of officers are deployed in a relatively small geographic area for the purpose of deterring and detecting criminal activity.

Saturday night special   an inexpensive cheaply made handgun. Because of their low cost, Saturday night specials are favored by some criminals.

scaffold   a platform on which a condemned criminal is executed by *hanging*. Scaffolds typically have a trap door through which the condemned drop to their deaths.

scam   a dishonest or illegal strategy to deceive others for financial or material gain.

scammer   one who perpetrates a *scam*.

Scandinavian Research Council for Criminology   a collaboration of the justice ministries of Denmark, Finland, Iceland, Norway, and Sweden whose purpose is to advance criminological research and offer advice on criminological issues. The Council is headquartered in Aarhus, Denmark.

scanning electron microscope   an extremely powerful microscope that permits forensic scientists to examine minute samples of *evidence.*

Scared Straight   an inmate-run program that originated in the Rahway State Prison in New Jersey designed to acquaint youthful offenders with the grim realities of prison life. Scared Straight consisted of inmates, many of whom were serving life sentences for murder, who related graphic portrayals of physical and sexual assault and other forms of exploitation against young male inmates. Evaluations of such programs has shown that these had little effect on the subsequent criminal involvement of attendees. Scared Straight programs also violated the sight and sound requirements of federal juvenile justice legislation.

school crime   crime committed in and around a school. School crime has always been a problem, but it has become more visible with a series of particularly devastating incidents, including the *Columbine massacre* and the *Newtown massacre.* School officials have been criticized for downplaying the extent and seriousness of school crime and for failing to maintain or disclose crime-related statistics.

school resource officer (SRO)   a law enforcement officer assigned to work in and around schools. In addition to providing security, SROs typically mentor students and educate them about *substance abuse* and other issues.

school safety   umbrella term that includes any program or movement intended to ensure the protection of students, teachers, and staff from intentional harm. School safety programs can include the use of a *school resource officer* or metal detectors to prevent weapons from being brought into school.

school-to-prison pipeline   the hypothetical route youths travel by being pushed out of school and into the criminal justice system due to lack of necessary opportunities and support.

school violence   violence occurring in or around a school. See *Newtown Massacre, school crime.*

scientific misconduct   illegal or unethical research or publication practices during the performance of scientific research. Scientific misconduct includes the fabrication or falsification of data, plagiarism, as well as the falsification of credentials by those who apply for scientific positions. Under certain circumstances, individuals engaging in scientific misconduct can be charged with crimes like *fraud.* See *qui tam suit.*

Scotland Yard   the investigative division of the London metropolitan police. Scotland Yard has an international reputation for being a highly professional law enforcement organization.

search and seizure   the specialty in criminal law concerned with the ability of law enforcement to search criminal suspects and seize property as a result.

SEARCH Group, Inc.   since the early 1970s, an organization that promotes the development of criminal justice information systems in the United States. SEARCH, Inc., which is headquartered in Sacramento, California, receives funding from the *Bureau of Justice Statistics (BJS).*

search pattern   any of several systematic ways investigators comb crime scenes for *physical evidence.* Common search patterns include the spiral, wheel, grid and zone methods, each named for the means by which an investigator covers the *crime scene.*

search warrant   a legal document issued by a judge or magistrate that permits law enforcement officials to enter and search specified environs, such as a home or business, for the purpose of finding evidence, apprehending a suspect, or other legitimate purposes.

Second Amendment   an amendment of the Constitution of the United States which guarantees a citizen's *right to bear arms.* The *Second Amendment* has been used by the *National Rifle Association* and other organizations to assert individual rights to own and carry firearms. Opponents argue that when the Constitution's framers drafted its language, an armed citizenry was important to guarantee freedom. However, with well-developed armed forces these arguments may be less compelling. See *gun control, gun lobby.*

Second Amendment Foundation   an American nonprofit organization whose mission is to protect the provisions of the Second Amendment of the U.S. Constitution, particularly the *right to bear arms.*

secondary conflict   according to Thorsten Sellin's *culture conflict* theory, this is conflict that develops in society as it becomes more heterogeneous. As groups differentiate and develop their own distinct values, they eventually clash with one another. Compare with *primary conflict*. See *Sellin, Thorsten*.

secondary crime scene a location of investigatory interest other than the *primary crime scene*. Examples of secondary crime scenes can include the vehicle that a suspect used to escape or the motel or residence used after committing a crime.

secondary deviance in the *labeling perspective*, the deviance which results from the acceptance by an individual of his or her deviant identity. Compare with *primary deviance*.

secondary high explosive a high explosive that is susceptible to detonation after the initiation of a *primary high explosive*.

secondary prevention the prevention of delinquency in those who are at risk or who have already engaged in delinquent acts. Compare with *primary prevention, tertiary prevention*. See *crime prevention*.

secondary victim one who indirectly suffers as a result of a crime, such as a relative or friend of a victim of a serious crime. Many so-called secondary victims object to their experience being referred to as secondary, since many lost loved ones and consequently must live with grief and sorrow.

second degree murder murder that is not premeditated or committed in connection with another felony. Compare with *involuntary manslaughter, voluntary manslaughter*.

secretor one who secretes blood type A, B, or AB antigens in body fluids. Criminal suspects who are secretors set themselves apart from other offenders and therefore are more easily identified if body fluids are available at the *crime scene* or elsewhere.

secret police law enforcement authorities of totalitarian governments who investigate and accuse, often without *evidence*, and employ *enhanced interrogation techniques*. The KGB of the former Soviet Union is an example of secret police.

Secure Communities an initiative of Immigration and Customs Enforcement where information on criminal aliens is shared among law enforcement agencies without imposing new requirements.

securities fraud any form of deceptive practice designed to illegally profit from the manipulation of stocks or other securities. See *Boesky, Ivan, Milken, Michael*.

security alarm a device designed to emit an electronic or mechanical signal when an intrusion takes place.

security company a private firm that specializes in identifying and mitigating security threats for residential and commercial customers.

security guard a privately employed person whose job is to protect people and property. Pejoratively called rent-a-cops, security guards have limited authority. Compare with *police*.

security system a set of devices designed to prevent and detect unauthorized intrusion.

security threat group a group of individuals in a correctional facility who collectively pose a threat to the safety and security of staff and other inmates. Examples of security threat groups include gangs, those affiliated with the *Aryan Brotherhood* or other white extremists, as well as Black militants. Those affiliated with security threat groups often are segregated from one another by moving them to different institutions in order to minimize their ability to cause harm. Compare with *gang*.

sedition actions or speech that could prompt rebellion against a government or other authority. Compare with *treason*.

selective breath testing see random breath testing.

selective incapacitation the policy and practice of confining violent or chronic offenders based on statistical predictors, such as *prior record* and seriousness of the *instant offense*.

self-concept theory any criminological theory that emphasizes the importance of a youth's self-concept as a cause of, or an insulator against, delinquency. See *containment theory*.

self-control theory a criminological perspective which attributes criminal behavior to lack of self-control by individuals. See *General Theory of Crime*.

self-defense acting in a way to save one's self from death or injury.

self-esteem an individual's sense of his or her own self-worth. High self-esteem is thought by some to help insulate individuals from delinquency and crime.

self-fulfilling prophecy the tendency of a person to live up to the negative imputations made about him or her. Self-fulfilling prophecy is often used to describe delinquents who are labeled by the juvenile justice system. See *labeling perspective*.

self-immolation the practice of setting oneself on fire. Self-immolation is typically performed as an expression of religious or political protest.

self-incrimination verbal or written statements that serve to cast suspicion on the person making them.

self-injurious behavior behavior, such as cutting or burning oneself.

self-mutilation see self-injurious behavior.

self-reported crime crimes divulged in the course of a research study designed to measure, among other factors, the offenders' actual involvement in criminal activity. See *self-report study.*

self-report study a study where survey respondents or interviewees are asked to reveal the nature and extent to which they have engaged in crime or delinquency. Self-report studies, which can employ surveys or interviews, gained increasing popularity with the acknowledgment that traditional crime statistics, such as the *Uniform Crime Reports,* had severe limitations in their ability to accurately measure crime. Compare with *other report study, victimization survey.*

Sellin, Thorsten (1896–1994) a prominent criminologist of the 20th century. Sellin was educated at the University of Pennsylvania where he spent his entire academic career. He is perhaps best known for his monograph for the Social Science Research Council, *Culture Conflict and Crime* (1938), and his work with *Marvin E. Wolfgang* and Robert Figlio titled *Delinquency in a Birth Cohort* (1972). See *Pennsylvania School of Criminology.*

semiautomatic weapon a firearm designed to discharge and re-chamber a round with the single pull of the trigger. Semiautomatic weapons are particularly lethal due to the capacity of their magazines. Compare with *automatic weapon, revolver.*

sentence the legal consequence imposed by a court of law on a convicted *offender.* A sentence can consist of time in prison or jail, a fine, *probation, restitution* to the victim, other consequences or any combination of these.

sentence bargaining a form of *plea bargain* where the negotiations between the prosecutor and defense attorney focus on which sentence the prosecutor will recommend the judge impose. Compare with *charge bargaining.*

sentence disparity differences in sentences meted out to offenders convicted of like or similar offenses. Sentence disparities can be the result of discretion or legislative differences between jurisdictions.

sentencing the phase of court processes where the defendant's punishment is determined. In many felony cases, the presiding judge uses a *presentence investigation report* to assist in arriving at a more just *sentence.* Sentencing options include *prison* or *jail* terms, *probation, fines,* or other alternatives.

sentencing circle a form of restorative justice based on Native American tradition. See *restorative justice.*

S

sentencing commission   an official body of appointed members charged with examining sentencing laws and practices and recommending changes.

sentencing discount   a reduction in an offender's sentence in return for a guilty plea. See *plea negotiation, sentence bargaining*.

sentencing guidelines   legislatively determined criteria for the imposition of criminal sentences. Sentencing guidelines restrict judicial *discretion*.

Sentencing Project, The   a U.S. organization that promotes the development of sentencing alternatives. Based in Washington, D.C., The Sentencing Project has published numerous reports, many pointing out inequitable and discriminatory practices in the criminal justice system. The Sentencing Project was a major force behind the *Campaign for an Effective Crime Policy*.

sentry   a lookout.

sequestering   the practice of isolating a *jury* from any outside influences, such as family, friends, and electronic and print media. Sequestering rests on the assumption that in order to render a just and impartial verdict in a particular case, jurors must not be biased by information from the outside. Depending on the nature and complexity of the criminal case, jurors may be sequestered for weeks.

serial arson   the occurrence of two or more instances of *arson* by the same *perpetrator*. Like *serial murder*, serial arson is often sexually motivated. Compare with *fire setting*.

serial arsonist   one who engages in *serial arson*.

serial killer   one who commits *serial murder*. Also used to define a serial murderer. Infamous serial killers include *Theodore Bundy, Jeffrey Dahmer, John Wayne Gacy,* and *Wayne Williams*.

serial murder   a series of homicides committed by one or more offenders with a cooling off period between each one. Serial murder differs from most other forms of homicide in that it is sexually motivated. Compare with *mass murder, multiple murder, spree murder.* See *Bundy, Theodore, Dahmer, Jeffrey, Gacy, John Wayne,* and *Williams, Wayne*.

serial murderer   same as *serial killer*.

seriousness scaling   the use of *psychophysical scaling* techniques to assess the perceived seriousness of criminal, delinquent, or deviant acts. Seriousness scaling in criminology was introduced by *Thorsten Sellin* and *Marvin E. Wolfgang* in *The Measurement of Delinquency*. Two methods of seriousness scaling are category scaling and magnitude estimation scaling.

serology   the study of blood.

serotonin   a compound in the blood that also acts as a neurotransmitter. There are some who believe that serotonin levels influence criminal behavior. See *biocriminology*.

Serpico, Frank (1936–present)   a former New York City police officer who in the early 1970s exposed widespread *corruption* within the New York City Police Department. Serpico met considerable hostility from other officers, many of whom were involved in the corruption. His allegations led to the formation of the *Knapp Commission* and resulted in the conviction of numerous police officials at all levels. See *grass-eater*, *meat-eater*.

service bailiff   a *bailiff* employed by a court who serves summons, subpoenas, and other legal documents which must be delivered in person.

service revolver   see *service weapon*.

service weapon   the weapon, most often a handgun, a law enforcement officer carries while on the job. Compare with *off-duty weapon*.

severity   the punitiveness of a criminal sanction. Severity is one of several characteristics of criminal sentences focused on by adherents of the *classical school of criminology*. Compare with *celerity*, *certainty*.

sex crime   any of a wide variety of crimes involving sex as a motivating factor.

sex offender   one who engages in sex-related crimes, such as rape or child molestation. See *sex crime*, *sex offender notification*, *sex offender registration*.

sex offender notification   the process of formally notifying local residents that a convicted sex offender has moved into their community.

sex offender registration   the requirement by law that convicted sex offenders must register with local law enforcement agencies. Sex offender registration became popular in the 1990s in the wake of several highly publicized sex crimes. See *Megan's Law*.

sex offender registry   an official roster of individuals who have been convicted of sex offenses who must register by law. Sex offender registries are frequently accessible by the public.

sex offender residency restrictions   administrative or legal provisions specifying certain locations where convicted sex offenders may not reside.

sexploitation   term used to describe the commercial exploitation of sex through the production and distribution of sexually explicit materials. See *community standards*, *pornography*.

sex slave   a human captive, most often female, who is forced to engage in sexual relations with the captor.

sextortion   the sexual exploitation of another using nonphysical means, such as the threatened release of private information.

sex trade   the selling and transportation of humans, primarily female, for use in *prostitution*. See *human trafficking*.

sexual assault forensic examination (SAFE)   a forensic examination of victims of sexual assault by physicians, nurses, and other medical professionals. SAFE involves the collection of evidence that potentially can be used to identify and prosecute the *offender*. See *Sexual Assault Nurse Examiner (SANE)*.

Sexual Assault Nurse Examiner (SANE)   a nurse who is specially trained to examine and collect evidence from victims of sexual assault. SANEs collect biological and other potentially important evidence with a *rape kit*.

sexual deviance   any deviant sexual practice.

sexual homicide   a homicide where the perpetrator's motive is primarily sexual in nature.

sexual predator   an offender who chronically commits sexual offenses. The term sexual predator is also used by legislators to designate special legislation aimed at this type of person.

sexual psychopath   a *sex offender* considered predatory and beyond rehabilitation.

sexual sadist   a *sex offender* who derives pleasure from inflicting pain on his victims. See *sadism*.

shackles   metal restraints that close around the ankles or wrists, linked to chains. Compare with *handcuffs*, *leg irons*. See *belly chain*.

shakedown   the systematic, often unannounced, search for contraband in jails or prisons. Shakedowns routinely yield homemade weapons, drugs, and other contraband.

shall issue   a legal provision specifying that citizens who meet requirements should be issued a permit to carry concealed firearms. Compare with *may issue*.

shaming penalties   one of the consequences for committing a crime where the offender is subjected to the community's disapproval. Shaming penalties have their beginnings in Australia and New Zealand. See *disintegrative shaming*, *integrative shaming*.

shank   slang term for a homemade knife. Shanks can be fashioned from a variety of hard materials, such as spoons and bedsprings, and are common in prisons and other correctional facilities. Periodic shakedowns in prisons

S

invariably turn up a variety of contraband, including shanks. Inmates are adept at hiding shanks. Also referred to as a *shiv.*

sharp force trauma   an injury caused by a sharp weapon or implement that causes penetrating or incised wounds. Sharp force trauma includes stab wounds, incised wounds, slash wounds, and chop wounds. Compare with *blunt force trauma.*

Shaw, Clifford (1922–1991)   an American criminologist and figure of the ecological school of criminology, along with Henry D. McKay. Shaw took graduate training at the University of Chicago where he was influenced by sociologists Ernest Burgess and Robert Park. See *Chicago Area Project, Chicago School of Criminology, McKay, Henry D., Institute for Juvenile Research.*

Shawcross, Arthur (1945–2008)   a *serial killer* who murdered numerous women in New York in the 1970s and 1980s. Shawcross died in prison. See *serial murder.*

Sheldon, William (1898–1977)   a medically trained professor of psychology who developed a *typology* relating body types to temperament. See *constitutional theory, ectomorph, endomorph, mesomorph.*

shell games   games of chance where the player must guess the location of a pea under one of several shells. Unwary players do not realize that shell games are rigged and therefore impossible to win.

shelters   places where victims of *domestic violence* can temporarily live in safety and receive services. Shelters are most often operated by nonprofit organizations and do not disclose their location to the public for purposes of safety and security.

Shepard, Matthew (1976–1998)   a 21-year-old student from the University of Wyoming who was assaulted and left to die near Laramie, Wyoming because he was gay. See *gay bashing, hate crime.*

Sheppard, Sam (1923–1970)   an Ohio osteopath first convicted and then acquitted of the murder of his wife. Sheppard was represented by attorney *F. Lee Bailey.* The case attracted much attention and inspired the 1960s television series, The Fugitive.

sheriff   an elected official of a county whose duties include enforcing laws and serving legal notices. In the United States, sheriffs typically provide law enforcement services in unincorporated areas of a county. They also operate county jails where offenders serve short sentences and await conveyance to state prisons.

Sherman Antitrust Act   a law passed in 1890 to regulate corporate behavior. The Sherman Antitrust Act prohibits the development of monopolies and provides penalties for *price-fixing.*

S

shiv   same as *shank.*

shock-and-awe policing   law enforcement practices characterized by the overwhelming show and use of force.

shock incarceration   a sentence where the *offender* serves a brief period of confinement in prison followed by release on probation or parole. The thought underlying shock incarceration is that exposure to the harsh realities of prison life will shock the offender into remaining crime free. See *shock parole, shock probation.*

shock parole   *parole* following a brief period of confinement in prison designed to acquaint the offender with the realities of prison life. Offenders receiving shock parole typically spend more time in prison than those on *shock probation.*

shock probation   *probation* following a brief period of confinement in prison designed to shock the offender with the realities of prison life. Generally the sentencing judge imposes a term of incarceration. After a designated period of time, the inmate may then apply to the sentencing court to have a shock probation hearing. The inmate, either present or in absentia, is either granted or denied probation. If shock probation is granted, the inmate is released from prison under probation supervision for a specified period of time under certain conditions.

shock treatment   a medical procedure where a mental patient is given an electroconvulsive shock in order to restore improved mental health. Also referred to as electroconvulsive shock therapy.

shoeprint   the image of the sole and heel of a shoe left by an *offender* and used by law enforcement for identification.

shoplifting   theft in a retail store by concealing merchandise. See *booster girdle.*

shotgun   a *long gun* with a smooth bore that fires numerous pellets or a rifled slug. At short ranges, shotguns can inflict devastating injuries. See *riot gun.* Compare with *rifle.*

showup   British term for *lineup.*

shunning   the practice of avoiding those who have violated certain norms. Shunning is associated with the Amish culture. See *labeling perspective.*

siblicide   the killing of a sibling. Compare with *fratricide* and *sororicide.*

Siegel, Benjamin "Bugsy" (1906–1947)   an *organized crime* figure of the 1930s and 1940s. Siegel was instrumental in developing the gambling casino business in Las Vegas. He was shot and killed at the home of his girlfriend, Virginia Hill.

sight and sound separation the requirement that juveniles held in adult jails or lockups must be removed in such a way that the adults and juveniles cannot have contact. Sight and sound separation is based on the notion that any contact with adult inmates is detrimental to confined juveniles.

signature a distinguishing trademark of a *serial killer* or other offender evident at the crime scene. Examples of signatures include types of binding, the use of unusual knots, and peculiar injuries or positions of the body that are left by the offender. See *serial murder*.

signature analysis the forensic analysis of signatures in order to identify a *suspect*. See *signature*.

silencer a device, generally affixed to the muzzle end of a firearm, designed to suppress the sound. Silencers are legal to own if one pays the federal transfer tax. Also referred to as a *muffler* or suppressor.

Silkwood, Karen (1946–1974) a woman who took on the nuclear power plant where she worked for its disregard of health and safety standards. After gathering incriminating documents, she was en route to a meeting when she was killed in an auto accident. The documents disappeared from the accident scene. The nuclear plant, Kerr-McGee, subsequently was found guilty of contaminating Silkwood and was ordered to pay her estate $10.5 million in damages. See *corporate crime*.

Simpson, O. J. (1947–present) a football and movie star who was accused of killing his ex-wife Nicole Simpson Brown and her friend Ronald Goldman in 1994. Simpson's initial flight from authorities and subsequent trial captured the attention of the world with its combination of high-priced defense attorneys, known as the *Dream Team,* and mistakes made by law enforcement authorities. Although Simpson was acquitted of the murder charges, many still believe he is guilty, especially in light of compelling *DNA testing* evidence. Simpson was later convicted of robbery and sentenced to prison.

Sing Sing an infamous New York state prison on the banks of the Hudson River near Ossining, New York. Sing Sing was the site of numerous executions using the *electric chair*, including those of *Julius and Ethel Rosenberg*.

situational crime prevention the prevention of crime by recognizing that offenders make rational choices to engage in crime and by removing or reducing opportunities for offending. An example of situational crime prevention is the placement of jewelry in locked display cases that only store clerks can open.

situational variable in criminological research, a variable that assesses changing circumstances, such as the dynamics of an interpersonal exchange or an individual's financial condition.

Sixth Amendment   the amendment of the U.S. Constitution that guarantees among other rights: the right to trial by jury, a speedy trial, and effective counsel.

skel   a slang term used by some law enforcement officers for *offender*.

skeletal remains   the skeletal structure remaining after destruction or decomposition of soft tissue and organs of a body.

sketch artist   an artist who is capable of sketching renditions of suspects based on the recollections of witnesses. See *forensic artist, Identikit.*

skid mark   a mark made on pavement by the tires of a motor vehicle. Skid marks can be analyzed by investigators to determine details of, and responsibility for, motor vehicle accidents.

skimming   to take money off the top of financial proceeds or hide profits, often to avoid the payment of taxes. See *white-collar crime.*

skinhead   a member of a white *gang* who wears closely cropped hair and tattoos and is violent against members of certain racial and ethnic minorities. Skinheads subscribe to neo-Nazi beliefs.

slander   spoken statements that negatively affect the name or reputation of another. Compare with *libel.*

slap jack   a leather lead-filled device using for hitting. Compare with *blackjack.*

slavery   the practice of keeping humans in bondage against their will. See *abduction, human trafficking, white slavery.*

sleuth   a *detective*, either professional or amateur.

smack   slang for *heroin.*

small arms   firearms capable of being carried. These include the *handgun* and *long gun.*

SMART   abbreviation for *Office of Sex Offender Sentencing, Monitoring, Apprehending, Registering, and Tracking.*

Smart, Elizabeth (1987–present)   a woman who at age 14 in 2002 was abducted and held for nine months before she was rescued. While in captivity, Smart was repeatedly raped and threatened. Since her rescue she has become an activist and journalist.

Smith, Susan (1971-present)   a woman who murdered her two children by rolling the family car into a lake in South Carolina. Smith, who claimed a carjacker had taken her car with the children still in it, made tearful pleas on television for the safe return of her children. She was convicted of two counts of first degree murder and was sentenced to life imprisonment.

Smith & Wesson since the 1800s, a major manufacturer of firearms, especially handguns. Smith & Wesson, along with other gun makers, were the subject of lawsuits in the 1990s for manufacturing dangerous consumer products.

smuggler one who engages in *smuggling*.

smuggling the surreptitious transfer of goods, often for the purpose of avoiding the payment of tariffs or transporting illegal drugs. See *Customs and Border Patrol, United States*.

sneak-and-peak warrant a *search warrant* that permits law enforcement officer to enter premises without permission and search without seizing property or *evidence*.

sniper a marksman capable of shooting targets at long distances. See *Whitman, Charles*.

snitch term used to describe an *informant*. See *confidential informant*.

snuff film a film where a person is murdered on camera. For many years, it was presumed that snuff films were urban legends when people referred to them because no one had ever actually seen one.

sobriety checkpoint a temporary station on a roadway where law enforcement officers stop and assess the sobriety of motorists in order to reduce drunk driving.

social bond theory the criminological notion that the strength of a person's bonds to society affect the likelihood of becoming involved in delinquency and crime. Social bond theory is most closely associated with criminologist Travis Hirschi. See *control theory*.

social capital collective assets in a community that increase the likelihood of positive social adjustment, and conversely help insulate residents from *criminogenic factors*.

social class the social and economic stratum to which a person belongs.

social contract the unwritten agreement between self-interested individuals and the government where they consent to restrict their pursuit of self-interest in pursuit of peace. In return, the government agrees to protect the individual's rights and freedoms.

social control both formal and informal means of ensuring compliance with laws, rules, and norms.

social control theory any one of several perspectives in criminology which emphasize crime as a consequence of uncontrolled hedonistic individuals. See *General Theory of Crime*.

social defense a nonviolent nonmilitary response to aggression. Social defense, which favors demonstrations and boycotts over the use of weapons and force, applies to both domestic and foreign aggression.

Social Development Research Group a research group affiliated with the University of Washington whose mission includes conducting research on healthy behavior and positive social development and developing effective interventions. The Social Development Research Group is headquartered in Seattle, Washington.

social disorganization the myriad of problems besetting certain urban neighborhoods, including crime and disorder.

social ecology the field concerned with the relationship between humans and their environment.

social inquiry report in Britain, a *presentence investigation report*. Also known as social enquiry report.

social justice equity and fairness in all spheres of life, including, but not limited to economic opportunity, gender, freedom of expression, concern over natural resources, and human rights.

social learning theory a criminological theory advanced by Ronald Akers and Robert Burgess which asserts that criminal behavior is learned through various forms of reinforcement. Also *learning theory*.

socially redeeming value regarding potentially pornographic material, the concept that it has something positive to offer society. See *pornography*.

social network analysis a set of techniques that permit the analysis of interdependent relationships among individuals and how these influence individual and social behavior. Social network analysis has been used to examine the interrelationships among gang-involved youths.

social norms the customary rules that govern behavior in groups and societies.

social pathology behavior in society which has negative consequences, such as crime and delinquency.

social process theory a theory in criminology that asserts that crime is a consequence of socialization. An example of a social process theory is *differential association*.

Social Science Research Council a nonprofit organization that explores important public issues through social science research. The Social Science Research Council issues publications and funds social research, including studies of crime.

social therapeutic institutions in Europe, facilities for convicted offenders designed to emphasize treatment over punishment. Incorporated into penal codes, social therapeutic institutions are criticized by conservatives for not being retributive enough and by liberals for perpetuating the medical model of corrections. See *therapeutic jurisprudence*.

Society for Laws Against Molesters (SLAM) an organization committed to promoting tougher legislation to punish child molesters.

socio legal studies the field of study that focuses on how the law and social institutions influence one another, as well as how the law influences the everyday lives of people.

sociology of law the empirical study of law and legal institutions as instruments of social change.

sociopath an individual who has no conscience, does not profit from experience, and has little regard for the rights or feelings of others. See *antisocial personality disorder, psychopath*.

Soderman, Harry (1902–1956) a Swedish police officer and pioneer in *criminalistics*.

sodomy sexual intercourse with a member of the same sex or with an animal. Sodomy also refers to anal or oral sex between humans.

software piracy the illegal acquisition and reproduction or use of licensed computer software.

solicitor in Britain, a member of the legal profession qualified to advise clients and provide instruction to barristers. Compare with *barrister, counselor*.

solitary confinement usually enforced as a punishment or for the individual's protection, the isolation of an inmate from others. See *hole*.

somatotypes body types that are believed to have a relationship to the individual's personality. See *ectomorph, endomorph, mesomorph, Sheldon, William*.

Son of Sam in the 1970s, the pseudonym used by *David Berkowitz*, the infamous .44 caliber killer who shot and killed men and women he found at lovers' lanes and other isolated locations.

Son of Sam laws any state or federal laws designed to prohibit convicted criminals from profiting from their crimes by way of lucrative book or movie deals, or other means of commercial exploitation. These laws take their name from the *Son of Sam* killings by *David Berkowitz*, who attempted to sell his story to a major book publisher. New York was the first state to enact such a law. These laws have been controversial because they conflict with the *First Amendment*.

S

SORNA abbreviation for *Sex Offender Registration and Notification Act.*

sororicide the killing of a sister by a sibling.

soul murder term used to describe the effects of sexual abuse on a child. See *post-traumatic stress disorder.*

Sourcebook of Criminal Justice Statistics since 1973, an annual compilation of a wide variety of crime- and justice-related statistics. The Sourcebook of Criminal Justice Statistics includes data on arrests, reported offenses, victimization, citizen surveys, and correctional populations. Funded by the *Bureau of Justice Statistics*, the Sourcebook of Criminal Justice Statistics is prepared by the Hindelang Criminal Justice Research Center at the University at Albany, State University of New York.

Southern Police Institute a law enforcement education and training center affiliated with the Department of Justice Administration at the University of Louisville in Kentucky.

Southern Poverty Law Center a legal organization located in Montgomery, Alabama that tracks white supremacist and other hate groups and engages in litigation against them. The Southern Poverty Law Center is a nonprofit organization and is funded by contributions.

southern subculture of violence a culture in the southern part of the United States characterized by a preference for violence in response to insults or threats. See *reactive aggression.*

souvenirs in a *serial murder*, an object taken by the killer from the victim as a memento of the crime. Souvenirs can be anything from the victim's personal possessions to body parts. The identification of souvenirs can help law enforcement officials make linkages among related homicides.

Spanish Society for Criminological Research an organization based in Spain that promotes criminological research and its application to national and international issues.

spatial analysis the statistical analysis of crime-related patterns in geographic areas, such as those depicted on maps.

specialty court a court designed to address a specific social issue or offender problem, such as domestic violence, drugs, guns, and mental health. Proponents argue that specialty courts are an improvement over traditional courts in that they better meet the needs of the offender and society. Those opposed to specialty courts believe that traditional courts can adequately handle a variety of diverse cases. See *domestic violence court, drug court, gun court, mental health court.*

specification   a circumstance of a crime that carries an additional mandatory term of confinement for those convicted. An example of a specification is the use of a firearm during the commission of a robbery. Aside from the penalty the robbery carries, the law also specifies that the use of a gun results in an add-on of additional years to the sentence.

specific deterrence   the type of *deterrence* directed toward individuals who have already engaged in crime. Locking up an offender is an example of specific deterrence. Also referred to as *individual deterrence*. Compare with *general deterrence*.

Speck, Richard (1941–1991)   an infamous mass murderer convicted of stabbing and strangling eight student nurses in Chicago in 1966. Though convicted and sentenced to death, Speck was spared execution by the 1972 prohibition against the death penalty. Years after his conviction, Speck once again made news by his participation in videotaped homosexual acts while in prison. See *mass murder*.

spectator violence   violence perpetrated by those who attend sports events. Spectator violence by hooligans at European soccer games has resulted in numerous deaths and injuries. Much of the problem can be attributed to the excessive consumption of *alcohol*. Compare with *hooliganism*.

speedy trial   the right of a defendant to have their pending criminal case brought to trial within a specified period of time. The defendant may waive this right, permitting the criminal justice system to take longer to process the case. See *Sixth Amendment*.

split sentence   a sentence given in a criminal court which consists of a term or confinement coupled with conditional release, such as probation. Split sentences combine the *incapacitation* associated with confinement with the opportunities presented by *community-based sanctions*.

sports violence   the violence that often accompanies certain forms of athletics. Boxing is perhaps the most controversial sport where violence is not a by-product, but an expected part of the action. Certain groups of medical professionals, including the American Medical Association, have spoken out against sports violence.

spousal abuse   the psychological or physical mistreatment of one spouse by the other. Compare with *domestic violence*.

spree murder   a series of homicides committed within a short period of time, sometimes in conjunction with other felonies, such as robbery or sexual assaults, without any intervening cooling off period. A notorious example of spree murders are those of *Charles Starkweather* who in 1956 killed ten people over eight days. Spree murder is a form of *multiple murder*. Compare with *mass murder, serial murder, Starkweather syndrome*.

spuriousness   an apparent, but false relationship between two variables. The classic example of a spurious relationship regarding crime is the relationship between ice cream sales and crime. As ice cream sales go up, so does crime. This might lead some to conclude that ice cream sales cause crime. However, the relationship is spurious because both ice cream sales and crime are related to a third variable, hot weather.

spy   a person who surreptitiously collects intelligence on the activities of others. See *espionage, Walker spy ring*.

staged crime scene   the arrangement of a corpse, weapon, furniture, or other *evidence* to mislead investigators about how a crime took place. Investigators are adept at identifying a staged *crime scene*.

stakeout   the covert surveillance of person or location to gather *evidence*.

stalking   the often undetected following of one person by another for the purpose of harassment or with the intent to do bodily harm. Stalking gained national attention in the 1980s and 1990s when several Hollywood celebrities became victims. Many states have enacted legislation designed to punish stalkers and protect their victims.

Standard Ammunition File   a reference collection that catalogues standard civilian and military ammunition.

Standardized Program Evaluation Protocol   a method for assessing the extent that juvenile programs reduce recidivism of participants.

stand your ground laws   state laws permitting citizens to not back down from threats of force or bodily injury. See *Zimmerman, George*.

Stanford Prison Experiment   a controversial social psychology experiment conducted by Stanford University professor Philip Zimbardo and his students. Some students played the roles of inmates, others played the roles of guards.

Starkweather, Charles (1938–1959)   a young man who with teenage girlfriend Carol Ann Fugate went on an infamous murderous eight-day rampage throughout the Midwest which left ten people dead. Starkweather was executed in 1959. See *multiple murder, spree murder*.

Starkweather syndrome   the assertion in criminology that offenders who engage in *spree murder* are more likely to be more violent and versatile than other kinds of murderers. This syndrome takes its name from the spree killer *Charles Starkweather*.

starring   the star-shaped rupture of human flesh caused by the pressure of exploding gasses during a contact gunshot wound. Also referred to as a *stellate*. See *contact wound, entrance wound*.

state administering agency   in U.S. Department of Justice nomenclature, a state agency that administers federal formula or block grant programs, such as those of the *Bureau of Justice Assistance* or the *Office of Juvenile Justice and Delinquency Prevention*. Many, but not all state administering agencies also serve as state planning agencies. See *state planning agency*.

state crime   crimes of commission or omission by the state, specifically by governmental authorities.

State Justice Institute (SJI)   a private, nonprofit organization created by Congress in 1984 to foster improvements in state courts. The SJI periodically awards competitive grants to organizations and governmental agencies to address court-related problems or to undertake innovative practices. Projects sponsored by SJI include those related to training for judges and court personnel, substance abuse, and the application of technology. The SJI is headquartered in Alexandria, VA.

state planning agency   a unit of state government officially charged with overseeing criminal justice planning and policy development throughout the state. Many state planning agencies were created in the late 1960s in order to administer the federal funds made available through the *Law Enforcement Assistance Administration*. Some state planning agencies are also *state administering agencies*. See *National Criminal Justice Association*.

state police   a law enforcement agency that has broad legal authority to investigate violations of criminal and traffic laws anywhere within the state. Home rule states do not have state police and instead rely on local law enforcement authorities like sheriffs for enforcement.

state prison   a correctional facility operated by state authorities for the confinement of convicted felons. Compare with *federal prison*.

stationary killer   a *serial killer* who commits crimes within a restricted geographical area. Compare with *nomadic killer*. See *serial murder*.

stationhouse release   the release of an offender right after arrest without the filing of formal charges. Stationhouse release is an example of law enforcement's use of *discretion*. Compare with *diversion*.

statistical analysis centers (SACs)   an organizational unit in most U.S. states and territories formed for the purpose of collecting, analyzing, maintaining, and disseminating various criminal justice data. In addition to serving as repositories for crime and justice data, many SACs undertake

S

research and evaluation projects. See *Justice Research and Statistics Association*.

statistical power   the ability of a study to detect an effect.

statistical prediction   the use of statistical variables to predict the likelihood of some outcome. In criminological research, the outcome of interest might be offender *recidivism*. Compare with *clinical prediction*.

Statistics Canada   the national agency responsible for collecting and maintaining criminal justice statistics in Canada.

status offender   a minor whose offense would not be illegal if he or she were an adult. Status offenders include those charged with *runaway* or being *incorrigible*. The federal *Office of Juvenile Justice and Delinquency Prevention (OJJDP)* has made a commitment to force states not to detain status offenders in secure facilities, especially in close proximity to adult offenders. Compare with *delinquent*.

status offense   an *offense* committed by a juvenile which would not be illegal if the person were an adult. Examples of status offenses are running away from home and *truancy*. See *runaway, unruly*.

statute   a law.

statute of limitations   a law which specifies time limits for prosecuting specific crimes. Typically, there is no statute of limitations for the crime of *murder*, but there is for lesser crimes.

statutory rape   sexual intercourse with a person under the age of consent. Statutory rape, while defined as illegal, often involves sex between consenting individuals. In most jurisdictions, the age difference between the *offender* and *victim* is significant in defining the act as statutory rape.

stay of adjudication   the avoidance of conviction by agreeing to certain court-imposed conditions. On meeting the specified requirements, the court dismisses the case. A stay of adjudication keeps the defendant from a guilty verdict.

stay of execution   an order suspending the execution of a court order or judgment. Execution in this context does not refer to the *death penalty*.

stellate   a star-shaped wound caused by a contact gunshot wound to the head or body. Gases expand under the skin, causing a ragged wound resembling a star or cross. See *contact wound, entrance wound, starring*.

stereomicroscope   a microscope which allows the side-by-side examination of two specimens. Stereomicroscopes are used to examine two bullets for comparison.

stick-up boy   a small-time armed robber. See *robbery*.

stigma an undesirable characteristic or reputation that follows youths, those with mental illness, or others processed by agents of social control. Stigma can reduce future opportunities for the labeled person. See *labeling perspective, radical non-intervention.*

stimulant a drug that creates a sense of euphoria in the user. Other common effects of stimulants include sleeplessness and loss of appetite. Amphetamines are an example of a stimulant.

sting operation an organized secret effort by law enforcement authorities to investigate and arrest those suspected of engaging in professional or organized criminal activity.

stippling same as *tattooing.*

Stockholm Prize in Criminology a prize awarded annually by the Swedish Ministry of Justice to one or more individuals for outstanding achievements in criminology.

Stockholm syndrome a phenomenon which occurs when a hostage begins to identify with and grow sympathetic toward their captor and antagonistic toward the authorities. The syndrome takes its name after the city where a female bank robbery *hostage* in 1973 became emotionally attached to one of her captors. One of the most famous cases of the Stockholm syndrome was that of newspaper heiress *Patty Hearst*, kidnapped by members of the *Symbionese Liberation Army* in 1974.

stocks a wooden structure used in colonial times to publicly punish and humiliate people for minor offenses. Compare with *pillory.*

stoner gangs gangs of white youths who identify with heavy metal music, punk associations, and sometimes satanic cults. Some stoner gangs have affiliated with white extremists. The term stoner derives from a person being stoned on drugs. See *White Aryan Resistance.*

S.T.O.P. abbreviation for *Stop Turning Out Prisoners.*

stop-and-frisk the law enforcement practice of stopping and patting down a person, usually one suspected of carrying a weapon, illegal drugs, or other contraband.

Stop Stick a long triangular box filled with sharp steel spikes designed to stop the motor vehicle of a fleeing suspect by puncturing the tires. Law enforcement officers lie in wait for the suspect vehicle and then pull a rope, drawing the Stop Stick into the vehicle's path. Once the tires pass over the Stop Stick, hollow spikes penetrate and insert in the tire, causing deflation and assisting pursuing officers to apprehend the suspect.

S

Stop Turning Out Prisoners (STOP)   a movement that originated in Florida devoted to stemming the early release of violent prison inmates. STOP proposes legislation and undertakes publicity campaigns.

strain theory   any theoretical perspective in criminology which emphasizes various types of strains on individuals that in turn make them more likely to engage in crime or other forms of deviant behavior. See *anomie theory, General Strain Theory*.

STR analysis   abbreviation for short tandem repeats, a method of analyzing *DNA* that is better at discriminating between two individuals. See *DNA testing*.

stranger abduction   the *abduction* of an individual by someone unknown to the individual. Stranger abductions, while capable of creating fear in parents for their children's safety, comprise a relatively small proportion of all abductions.

stranger danger   a phrase used to educate and warn children about the potential threat posed by predatory strangers.

stranger rape   the rape of an individual by someone unknown to the victim. Compare with *acquaintance rape*.

strangling   the act of compressing the throat of a person, thereby cutting off the air supply, eventually resulting in serious injury or death. See *garrote*.

straw purchase   the purchase of a firearm by a qualified buyer for another individual who intends to illegally possess, use, sell, or transfer it. Straw purchases that circumvent gun laws account for many guns getting into the hands of criminals. See *gun control, Operation Fast and Furious*.

street crime   conventional unsophisticated crime, such as robbery and theft. Street crimes, which do not require special skills or education to commit, comprise the majority of crime in the United States. Compare with *upper-world crime, white-collar crime*.

street gang   a *gang* composed primarily of juveniles and young men who engage in criminal activity.

strict liability   a theory of close adherence to the letter of the law, even in the face of ignorance that a violation has occurred. According to strict liability, a person who unknowingly buys stolen goods is criminally liable even though he was not aware that the merchandise was "hot." Compare with *vicarious liability*.

strip search   the search of an accused suspect or convicted offender that involves the removal of clothes and the search for weapons and other contraband over the entire person, often including body cavities. Strip searches are necessary because some criminals are adept at hiding contraband.

Stroud, Robert (1890–1963) a convicted federal prisoner who became known as the Birdman of *Alcatraz*. Stroud was an extremely violent offender who had to be segregated from other prisoners.

structural of or relating to the social structure and broad social forces in society, such as socioeconomic status and stratification.

structural criminology a branch of *criminology* that emphasizes the role of structural factors in the genesis and transmission of crime, such as poverty, economic disadvantage, racism, and other societal factors. Structural criminology plays down the role of individual factors like personality. See *power-control theory*.

structural equation modeling a set of methods used to specify the relationship and causal order of latent variables in criminological research.

structural Marxism a more scientific Marxism that argues that the state and its institutions promote capitalism and the interests of the ruling class. Compare with *instrumental Marxism*.

structural theory any theory in criminology that emphasizes the role of structures in society, such as class and institutions.

student threats threats made by students against teachers, other students, or the school. See *school crime*.

stun gun an electrical handheld device that permits the user to render another person temporarily immobile through the application of a nonfatal, high-voltage electrical charge. Compare with *TASER*.

subculture a smaller cultural group with beliefs, *norms*, practices, and rituals that are different from and sometimes at odds with the larger culture. Examples of subcultures are *gangs* and *organized crime*. See *subculture of violence*.

subculture of violence a term coined by criminologists *Marvin E. Wolfgang* and *Franco Ferracuti* to describe a *subculture* in society whose *norms* support violence as an acceptable way of life. An example of a subculture of violence is that found in the southern United States where affronts to one's honor are typically met with a violent response. Gangs, also subcultures of violence, operate under different norms from those of larger society.

subculture theory any criminological theory that focuses on the distinct characteristics or criminogenic effects of a subculture.

subpoena a writ issued by a court or other government body compelling a person to testify or produce *evidence*.

subpoena duces tecum a *subpoena* that includes an order to produce certain documents or other *evidence*.

S

sub rosa indictment   a secret *indictment*. Sub rosa indictments often are used when officials fear that any publicity will compromise their ability to successfully apprehend and prosecute the party in question.

substance abuse   the use of drugs or alcohol that poses a risk of health, legal, occupational, or other problems.

Substance Abuse and Mental Health Services Administration (SAMHSA)   a division of the U.S. Department of Health and Human Services responsible for promoting programs and providing funding for substance abuse and mental health issues.

substantial capacity test   a test of insanity which asks whether the accused had the substantial capacity to appreciate the wrongfulness of the conduct in question and, if so, to control that behavior. See *Durham rule*.

subterranean values   the values of deviant subcultures which differ from those of mainstream society.

suffocation   a condition caused by the interruption of the flow of air for breathing. Suffocation is a common cause of death in homicides, especially those of infants. See *burking*.

suicide   the intentional taking of one's own life. Suicide is actually a form of *homicide*, and in most jurisdictions is investigated as such. See *inmate suicide, self-injurious behavior*.

suicide bridge   a high bridge notorious as a place where suicidal individuals jump to their deaths. Examples include the Golden Gate Bridge in San Francisco Bay and the Prince Edward Viaduct in Toronto, Canada. See *suicide*.

suicide by cop   the intentional effort by a person to force a police officer to shoot. Some suspects refuse to surrender their weapons and even charge the police, leaving officers no alternative except to use *deadly force*. See *suicide*.

suicide cluster   a group of suicides that appear to be related geographically, temporally, or for some other reason.

summary executions   executions carried out quickly without due process of law for the condemned.

summons   an order to appear in court. Compare with *subpoena*.

superglue fuming   see *fuming*.

superior court   a court where felony cases are heard.

supermax prison   a maximum security prison or unit designed to securely house the most dangerous convicted prisoners. Because supermax prisons often segregate offenders and keep them in isolation, it is argued that this prison system constitutes *cruel and unusual punishment.*

super predator   term used to describe youthful offenders capable of greater amounts of more serious criminal behavior than their predecessors. Some criminologists predicted that in the early 21st century, super predators would proliferate, creating a wave of violent crime.

supervised release   the release of a convicted offender from imprisonment followed by a period of supervision requiring certain conditions. Compare with *parole.*

Supplementary Homicide Reports (SHR)   adjunct reports on criminal homicide produced routinely by local law enforcement agencies as part of their participation in the *Uniform Crime Reports* program. Data collected through SHR include the offender's age, sex, and race, the victim's age, sex, and race, the circumstances of the offense, the offender-victim relationship, and the type of weapon. SHR has largely been replaced by the *National Incident Based Reporting System (NIBRS).*

supply reduction   a drug control policy that focuses on restricting the supply of drugs available for consumption. Compare with *demand reduction.*

supreme court   in courts of appeals, the highest court to which an appeal can ascend. Rulings of the U.S. Supreme Court are final.

Sureños   a large Hispanic street *gang.*

surety   one who assumes responsibility for the appearance in court of another.

surety bond   a promise to pay a certain amount of money to a party in the event someone fails to meet an obligation.

surveillance   close watch over something, such as a criminal suspect. Known informally as *stakeout*, surveillance by law enforcement is surreptitious.

surveillance equipment   mechanical or electronic devices designed and used to monitor others. Examples are fiber-optic cameras and small listening devices, also referred to as bugs.

survival analysis   a statistical technique used to measure the time between the onset of a disease and a terminal outcome. Survival analysis was originally developed to study the survival of patients with diseases, such as cancer, and has been employed by criminologists to measure how long offenders remain crime free after they have undergone various forms of treatment.

survivalist   a radical, often heavily armed individual who fears a takeover by foreign or domestic governments. Survivalists stockpile provisions in the event they have to retreat from mainstream society and defend their freedoms.

survivor   term used to describe the living victims of crimes that often resulted in serious injury or death, especially violent offenses like domestic violence. The meaning of the term has been broadened to include the family and close friends of the victims of violent crimes, including homicide, who continue to suffer long after the commission of the crime. See *Children of Murdered Parents, Parents of Murdered Children, secondary victim.*

suspect   a person believed to be responsible for a crime. Compare with *accused, arrestee.*

suspected item disposal   the safe disposal of an item that could be potentially harmful, such as a suspected bomb. See *bomb squad.*

suspense novel   a fictional book where the main character is placed in jeopardy, creating suspense on the part of the reader. Compare with *crime novel, mystery, thriller.*

Sutherland, Edwin H. (1880–1951)   a leading figure in 20th-century American criminology, considered by many as the father of modern criminology. Sutherland took his graduate training at the University of Chicago, and taught at several universities before moving to Indiana University for the remainder of his career. His most notable writings include *The Professional Thief, White-Collar Crime,* and *Criminology,* a textbook coauthored with *Donald Cressey,* which was published in numerous editions over the course of 40 years.

Sutton, Willie (1901–1980)   an infamous American bank robber of the 20th century.

Swango, Michael (1954-present)   a physician and *serial killer* who murdered an undetermined number of patients by poisoning them. See *angel of death, serial murder.*

S.W.A.T.   short for Special Weapons and Tactics. A specialized squad of law enforcement officers whose job is to handle high-risk operations in progress, including hostage negotiations, *high-risk entry,* and violent felonies. SWAT officers typically are trained in repelling and scaling as well as the use of a variety of weapons.

swiftness   see *celerity.*

swindling   the cheating of a person out of money or possessions.

switchblade   a spring-loaded knife designed to open quickly with the push of a button. Under federal law, switchblades are illegal to carry in the United States. Compare with *ballistic knife.*

Symbionese Liberation Army   an American left-wing group that committed a number or crimes, including murder, armed robbery, and the 1974 kidnapping of heiress *Patty Hearst.*

symbolic interactionism   a social psychological perspective that emphasizes subjective impressions and interpretations in human interaction. Symbolic interaction, which arose out of the work of early social psychologists Charles Horton Cooley and George Herbert Mead, helped lay the theoretical groundwork for the *labeling perspective* in criminology.

Synanon   a well-known drug rehabilitation program that disbanded in 1989.

syndicate   a large criminal organization. Like the term *mob*, syndicate is frequently used for *organized crime*.

systematic check forger   a forger who purposely engages in forgery as a business. Compare with *naive check forger*.

systematic review   a compilation of in-depth evaluation information about a program. See *Campbell Collaboration, Cochrane Collaboration*.

taboo   a socially forbidden activity or custom.

tagging   the practice by *gang* members or others of spray painting *graffiti*, often to mark their territory. Also, a British term for *electronic monitoring*.

Tailhook scandal   an incident involving U.S. Navy personnel where male officers assaulted 26 women at a convention of the Tailhook Association in Las Vegas in 1991. While the officers' behavior was cast as sexual harassment, it could have been charged as sexual assault. A number of those involved were formally disciplined.

tailing   the practice of following another person, generally in a motor vehicle, in order to maintain visual contact or to determine their destination. Implicit in tailing is that the *surveillance* should not detected.

takeover robbery   a *robbery* where the offenders take over the establishment under attack.

tampering with evidence   the illegal removal, destruction, or alteration of *evidence* in a criminal case.

Tarasoff decision   court decision requiring that mental health professionals be proactive in taking measures to protect individuals who are being threatened with bodily harm by a patient.

Tarde, Gabriel (1843–1904)   a French magistrate and sociologist who developed the forerunner to learning theory in criminology. Tarde, who rejected the work of *Cesare Lombroso* and other constitutional theorists of the day, argued that people become criminals by imitating the behavior of others. His work influenced a number of later criminologists, including *Edwin H. Sutherland*. See *learning theory, social process theory*.

target removal   efforts to eliminate the potential object of a criminal's intentions. Target removal, which is consistent with the *routine activities approach* in criminology, is based on the notion that criminals cannot victimize what is unavailable. An example of target removal is not leaving valuables in a vehicle where they are visible. See *crime prevention*.

T

TASC   abbreviation for *Treatment Alternatives to Street Crime.*

Taser   an electronic weapon that works by discharging high voltage electrodes attached to long wires that when they penetrate human flesh, render the individual temporarily immobile. Originally designed for self-protection by law enforcement, security, and other well-meaning, innocent persons, Tasers have been used by criminals for committing crimes. Compare with *stun gun.*

tattooing   tattoo or pigment-like alterations to human skin caused by grains of gunpowder being driven into the skin. The presence of tattooing can reveal the distance between a shooter and the person shot. See *entrance wound, gunshot residue.*

tax evasion   the intentional and illegal nonpayment of taxes to the government. Tax evasion was the crime for which *Al Capone* was eventually convicted and imprisoned.

tax haven   a location, most often a country or possession, where income tax is low or nonexistent. Tax havens are used by those who want to avoid paying taxes in their country of residence.

taxonomy   a classification of something, such as criminal offenders or criminological theories.

tea room   a term used for public bathrooms such as those found in parks or roadside rest areas where homosexual activity takes place.

tea room trade   impersonal homosexual activity in public restrooms or other public places.

technical violation   an infraction of the rules of *probation* or *parole.* Technical violations, which are considered separate from new criminal charges, can include such infractions as moving without permission or associating with a *codefendant.*

teen court   a quasi-judicial process operated by teenaged youth for the purpose of settling disputes and learning about the justice system. See *specialty court.*

telemarketing fraud   the illegal solicitation of donations, products or services by telephone which results in undervalued or no merchandise for the purchaser. This type of scheme is sometimes used simply to get the credit card numbers of unwary customers.

telescoping   a narrowing of focus in a criminal case such that other potentially pertinent information or suspects are overlooked or disregarded. Compare with *linkage blindness.*

temporary restraining order   a *restraining order* that is in force for a limited period of time.

territorial killer a *serial killer* who commits his crimes within a circumscribed geographical area. See *geographic profiling, serial murder.*

terrorism the advocating or use of violence to coerce government action. Terrorism differs from other forms of crime in that it is most often motivated by political or religious considerations. Examples of terrorism include the hijacking and crashing of commercial jets into the World Trade Center and the Pentagon. Terrorism often differs from other forms of crime in its *method of operation.* See *Attack on America, domestic terrorism, international terrorism.*

terrorist one who engages in *terrorism. Timothy McVeigh* and *Osama bin Laden* are notorious terrorists of the late 20th and early 21st centuries. See *domestic terrorism, international terrorism.*

tertiary prevention the prevention of undesirable traits or behavior in those persons already afflicted with a condition or disease. The tertiary prevention of crime is more the control of future offending. For example, the tertiary prevention of violent crime might consist of *incapacitation* combined with intensive treatment. Compare with *primary prevention, secondary prevention.*

testimony information a person gives verbally in court that is presumed to be the truth.

Texas Rangers a division of the Texas Department of Public Safety with a long and storied history of law enforcement. Their current responsibilities comprise investigating major crimes, including serial crimes, public corruption, and officer involved shootings.

theft the intentional and illegal taking of property. See *grand theft, petty theft, theft by deception.*

theft by deception illegally gaining the property of another by using lies or other forms of deceit.

theoretical criminology the aspect of criminology that is concerned with the formal explanation of crime and the criminal justice system.

therapeutic community a treatment philosophy and related practice based on the segregation of clients and the use of a *token economy.* Research shows that therapeutic communities achieve positive outcomes for those addicted to substances of abuse. See *substance abuse.*

therapeutic jurisprudence an approach to the use of law as a potential therapeutic agent. Therapeutic justice rests on the assumption that the law has the power to effect dramatic social consequences, and that these consequences should be considered and studied. Of special interest in therapeutic jurisprudence is the law's impact on the psychological and emotional well-being of individuals.

thermic law of crime   Adolphe Quetelet's conception that crimes against persons tend to occur in warmer climates whereas property crimes are more likely to occur in colder climates. Empirical support exists for the thermic law of crime. See *Quetelet, Adolphe.*

thief   a person who commits *theft.*

thin blue line   term used to describe the fragile barrier law enforcement represents between a safe society and lawlessness.

thin layer chromatography   a technique used by forensic scientists to separate nonvolatile mixtures, such as those found in suspected drugs.

Thompson submachine gun   a .45-caliber submachine gun that was widely used by gangsters in the early part of the 20th century.

threat analysis   see *threat assessment.*

threat assessment   the analysis of the potential dangerousness of people and situations in order to mitigate risk.

threat assessment team   a team trained to perform a *threat assessment.*

three strikes and you're out   a penal philosophy policy expressed in state laws that imprisons for life offenders who attain their third conviction. Three strikes and you're out, a policy which began in California, has been controversial in that some offenders can in fact strike out for a relatively minor offense, such as shoplifting. Consequently, it may not further the ends of justice. It also has been criticized because there is little reason to believe that such a policy will have an appreciable effect on the amount of crime. See *tough on crime.*

thriller   a fast-paced *crime novel* in which the protagonist is in jeopardy. Compare with *mystery, suspense novel.*

throwaway   a weapon left by a law enforcement officer at the scene of the shooting of an unarmed suspect. The throwaway is intended to make the shooting appear justified.

ticket fixing   arranging for the cancellation of a traffic ticket through political connections.

ticket of leave   in colonial Australia, the official permission given to an offender for the purpose of establishing a residence and finding employment prior to the expiration of the sentence. The ticket of leave is said to be the precursor of present-day *parole.*

time series analysis   a statistical technique used to forecast future trends in crime rates, correctional populations, or other criminal justice applications.

Time series analysis examines trends over periods of time. Compare with *ARIMA*.

tipping point   the point in an epidemic, including a crime epidemic, at which the disease spreads rapidly throughout a *population*.

tire marks   surface impressions made by a vehicle's tires. When tire marks are found at a crime scene, investigators can use a combination of photography and casting to determine the make and model of a particular tire through laboratory analysis. Such castings can later be compared to suspect vehicles to determine their involvement in a crime.

To Catch a Predator   a television program where adult men who have scheduled sexual liaisons with juveniles, most often using the Internet, are exposed on camera and arrested. See *sting operation*.

toe tag   an identification tag affixed to the toe of a deceased person.

token economy   a means of exchange used by residents in a *therapeutic community* for behavior modification.

Toole, Ottis (1947–1996)   a drifter and *serial killer* who is suspected of having murdered *Adam Walsh*. See *Lucas, Henry Lee, serial murder*.

tool mark   any mark made by an implement that a criminal suspect may have used in the commission of a crime, such as a screwdriver. See *tool mark identification*.

tool mark identification   the analysis of scratches and other marks found in metal and other hard materials at crime scenes for the purpose of making linkages.

torture   the intentional infliction of pain upon another, often for the purpose of forcing the tortured person to reveal information. Alleged torture around the world is closely monitored by *Human Rights Watch*. Compare with *enhanced interrogation techniques*.

total institution   a setting where individuals are isolated from larger society. Total institutions are bureaucratic, rigid, and closely control the lives of their charges. A prison is an example of a total institution.

total station   a device for facilitating *forensic mapping* by electronically measuring distances.

tough on crime   a stance by politicians characterized by severe penalties. An example of being tough on crime is *three strikes and you're out*.

toxic criminals   a term given to corporations that create environmental disasters through the dumping or otherwise irresponsible handling of hazardous waste. See *corporate crime*.

toxicology   the study of poisons and other toxic substances.

toxicology report   a report that summarizes the types and amounts of drugs, poisons, or other substances in a deceased person.

trace evidence   physical *evidence* left behind as the result of a crime.

trace metal   small amounts of metals that can be found in human tissue. *Arsenic*, a common poison, is an example of a trace metal.

traffic accident reconstruction   an investigative specialty of analyzing traffic accidents to determine the sequence of events.

trafficker   one who engages in *trafficking*.

trafficking   the sale and distribution of illegal materials, including drugs, guns, and people.

training school   a residential school where delinquents receive vocational training.

trait theory   any criminological theory concerned with the traits of offenders.

trajectory   the path of a bullet or other projectile. See *ballistics*.

transcript   a written record of a criminal case proceedings.

transitional justice   justice that purports to correct the wrongs associated with *human rights violations*.

translational criminology   efforts to make criminological theory and research relevant to criminal justice policy and practice, particularly through the use of *evidence-based practice*.

transnational crime   *organized crime*, corruption, and other forms of criminal behavior whose influence spans the globe.

transportation   the movement of convicted criminals from one geographical location to another, generally to a more remote location. Australia was once populated with English convicts who found their way to that country through transportation.

transportation officer   a police officer or correctional officer charged with transporting prisoners from one location to another.

Transportation Security Administration (TSA)   the federal agency within the Department of Homeland Security whose purpose is to assess the risks and strengthen the security of transportation systems in the United States. TSA personnel are especially visible at airports.

treason   the act of violating one's allegiance to a country by attempting to overthrow or otherwise subvert the government. Treason is an extremely serious crime, in some cases punishable by death. See *espionage, Walker spy ring*.

Treatment Alternatives to Street Crime (TASC) a national program designed to provide *substance abuse* treatment for offenders involved in the criminal justice system.

trespassing the unauthorized presence of a person on the property of another, which is prohibited by law. Compare with *criminal trespass*.

Triad a collaboration consisting of a *sheriff*, a *police chief* and a senior citizen group in the community who join together to reduce criminal victimization and meet the needs of retired and other older persons. See *crime prevention*.

trial court process in which allegations are examined and decided. Due to the extensive use of *plea negotiation*, most criminal cases do not go to trial. See *bench trial, jury trial*.

trial by fire a *trial by ordeal* where the accused was exposed to flames.

trial by jury *jury trial* as an advance in the evolution toward *due process*. Compare with *trial by ordeal*.

trial by ordeal an ancient practice to determine guilt or innocence where the accused was subjected to painful and potentially deadly procedures. If the individuals survived, which they often did not, they were deemed innocent.

trial by water one form of *trial by ordeal* where the accused was submerged in water in order to establish guilt or innocence. Compare with *waterboarding*.

trial court the lower court where criminal cases are tried. See *trial*. Compare with *appeals court, supreme court*.

triangulation a method of obtaining measurements at a *crime scene*. Triangulation, often recorded on graph paper, is accomplished by choosing two points other than the location of the object in question. Measurements from each point to the object are taken and transferred to the sketch. Where the lines intersect on the sketch is the location of the object.

tribal justice justice administered by indigenous tribes.

Trojanowicz, Robert (1941–1994) late professor of criminal justice at Michigan State University and regarded by many as one of the founders of *community policing*. Trojanowicz, who developed the twelve principles for community policing, was also affiliated with the Kennedy School of Government at Harvard University.

trophy an item belonging to the victim that a *serial killer* keeps as a remembrance. See *serial murder*.

T

truancy   the unexcused or unauthorized absence of a student from school. In some cases, truancy can lead to intervention by juvenile court. See *status offense, unruly.*

truant   a youth who is absent from school without authorization. See *status offense.*

true bill   the indictment document returned by a grand jury stating a defendant has formally been accused of one or more offenses. Compare with *bill of information, indictment.*

true crime   a category of nonfiction that features actual crimes. An example of true crime is the book *In Cold Blood* by *Truman Capote.*

truncheon   a short, club-like weapon carried by a police officer.

trusty   a prison or jail inmate who is entrusted with duties or responsibilities requiring greater trust than that accorded other inmates. Typical trusty duties include chauffeuring automobiles for government officials and performing custodial duties.

trusty security   a level of security that permits greater freedom for the *trusty*. Compare with *close security, maximum security, minimum security.*

truth in sentencing   a phrase that conveys the intent to have convicted offenders serve the amount of jail or prison time to which they were sentenced. Expressed in numerous state laws enacted in the latter part of the 20th century, truth in sentencing as a philosophy was a punitive response to the perception of victims and others that offenders were serving too little of the sentences they were given in court.

Tsarnaev brothers   Chechen brothers who are alleged to have perpetrated the *Boston Marathon bombings* in 2013. Tamerlan was killed in a shootout with police and his brother Dzhokhar was later wounded and captured.

Tsarnaev, Dzhokhar (1993–present)   see *Tsarnaev brothers.*

Tsarnaev, Tamerlan (1986–2013)   see *Tsarnaev brothers.*

tunnel vision   a narrow focus on one or more potential criminal suspects, eliminating other suspects from serious consideration. Tunnel vision can lead to a *wrongful conviction.*

Turner Diaries, The   a book written by William Pierce under the pen name of Andrew McDonald and published in 1980. *The Turner Diaries* is an example of modern neo-Nazi propaganda, espousing a revolution against the U.S. government because of its protection and support of Jews and certain targeted minorities.

twin studies   any study employing identical twins in order to test for the inheritability of criminal tendencies. Twin studies enable researchers to look at the effects of environment on twins reared separately.

Tyburn tree   an infamous *gallows* erected outside London in the 19th century. Originally designed for single hangings, the Tyburn Tree was later modified to have three sides, each of which could accommodate several condemned offenders. See *hanging*.

Tylenol murders   murders in Chicago in 1982 caused by people taking cyanide-laced Tylenol they had purchased in area stores.

typology   a system of classifying things according to specified criteria. Typologies in criminology include those classifying types of offenders, crimes, victims, and theories of crime. Compare with *taxonomy*.

UCR abbreviation for *Uniform Crime Reports.*

Unabomber nickname given to the person or persons responsible for the bombings eventually connected to *Theodore Kaczynski.*

unauthorized use of a motor vehicle a criminal offense involving the use of a motor vehicle without permission, which is a less serious offense than theft of a motor vehicle. Compare with *auto theft, carjacking, joyriding.*

unconditional release the release of convicted offenders from prison and once they are released, they can no longer be imprisoned for the sentences they had served.

underboss in an organized crime family, the highest ranking official just below the *boss.* Compare with *consigliere.*

undercover the state of infiltrating and investigating while concealing one's true identity.

undercover officer a law enforcement officer who poses as a drug dealer or assumes another role in order to investigate or infiltrate criminal enterprises.

undercover operation a criminal investigation where law enforcement officers pose as members of criminal organizations or otherwise disguise their true identities in order to gather detailed information about the organization's inner workings. Undercover operations are common in narcotics investigations.

undercriminalization the tendency of law to not treat seriously enough socially harmful behaviors.

undersheriff an officer whose position is directly under the *sheriff.*

undocumented alien preferred alternative term for *illegal alien.*

unidentified suspect a possible criminal *offender* whose identity remains unknown.

Uniform Code of Military Justice the body of criminal laws and associated justice procedures governing the behavior of armed services personnel in the United States.

Uniform Crime Reports (UCR) a crime statistics program of the *Federal Bureau of Investigation (FBI)*. UCR has been the staple of national crime reporting since the 1930s. UCR collects summary-based information from law enforcement agencies on offenses reported, arrests made, and more detailed data on homicides. The UCR is gradually being replaced by the *National Incident Based Reporting System (NIBRS)*.

unintentional injury an injury that has an accidental cause. Compare with *intentional injury*.

Unione Corse a French organized criminal enterprise similar in nature to the *Mafia*.

United Bamboo a Taiwanese criminal organization. See *organized crime*.

United Nations Office on Drugs and Crime (UNODC) the office within the United Nations which engages in crime control activities, including advancing research, supporting policy, combating terrorism, promoting justice, and countering organized crime and corruption.

United States Sentencing Commission an independent agency within the U.S. government whose purposes include establishing sentencing policies and practices and advising Congress and the executive branch on crime control.

Universal Declaration of Human Rights a statement issued in 1948 by the United Nations that spells out the basic freedoms people of the world should enjoy. The Universal Declaration of Human Rights was developed in response to the atrocities committed during World War II.

Universal Latent Workstation computer software that enables latent print examiners to transfer, analyze, and compare fingerprints. See *latent prints*.

University of Chicago Crime Lab a group of scholars at the University of Chicago and other universities who conduct research on issues of crime and justice.

unlawful assembly the illegal gathering of persons in a public place.

unlawful entry entering a residence or business without the owner's permission. Compare with *burglary*.

unobtrusive measures research techniques which do not involve contacting, interviewing, or otherwise inconveniencing research subjects. Unobtrusive measures include content analysis, the analysis of archival data and certain observational techniques. The primary advantage of unobtrusive measures is that the researcher is less likely to affect the data collection process.

unreported crime crime not reported to law enforcement authorities. In addition to an amount of crime in general not reported, specific types of crime, such as sexual assaults, tend to go unreported. See *dark figure of crime*.

unruly   a disobedient youth whose behavior leads to juvenile court involvement. Also the charge for such misbehavior. Compare with *delinquent.*

unsub   abbreviation for *unidentified subject.*

uppers   slang for a stimulant drug like *amphetamines.* Compare with *downers.* See *substance abuse.*

upperworld crime   criminal behavior by those in the upper socioeconomic strata of society. Examples of upper world crime include *insider trading* and *embezzlement.* Compare with *corporate crime, white-collar crime.*

Urban Institute   a private organization that conducts research and offers consulting services on a wide variety of criminal justice issues, including prisoner reentry.

urinalysis   the chemical analysis of urine to check for the presence of drugs. When such a test yields a positive result or what is known in criminal justice vernacular as *dirty urine,* it is indicative of drug use. Urinalysis is commonly used for substance abusers on *probation* or in treatment. See *substance abuse.*

U.S. Attorney   one of the attorneys who works under the direction of the U.S. Attorney General. U.S. Attorneys prosecute federal crimes like bank robbery.

U.S. Bomb Data Center   a federal center that collects and maintains data on incidents involving *arson* and explosives. The U.S. Bomb Data Center receives information from a number of federal agencies.

U.S. Sentencing Commission   a federal agency whose purposes include establishing federal sentencing policies and practices and conducting research on federal criminal sentencing issues.

usual suspect   a person typically involved in or suspected of involvement in crime.

usury   lending money at an exorbitant interest rate. While usury generally is associated with unlawful behavior, practices such as those employed by check-cashing businesses and similar loan establishments skirt the definition of usury with their high lending rates and exploitation of low socioeconomic patrons. Compare with *loan sharking.*

utilitarianism   the position in philosophy that happiness of the individual or of society in general is the desired end. See *classical school of criminology, Bentham, Jeremy.*

uxoricide   the murder of a wife by the husband.

vagrancy not having a stable home or job or the offense for such a condition. Many of those charged with vagrancy are suffering from mental illness. See *homeless persons*, *public order crime*.

Valachi, Joseph (1903–1971) a member of an *organized crime* family who testified before the *McClellan Committee* about the inside workings of the American *Mafia*.

valley a depression in a *fingerprint* detail. Compare with *ridge*.

vandal one who engages in *vandalism*.

vandalism the malicious alteration or destruction of property, most often by juvenile offenders, often without obvious motive.

vanity hypothesis the hypothesis which states that women are more likely than men to commit suicide using methods that do not disfigure the head and face.

Vanzetti, Bartolomeo (1888–1927) see *Sacco and Vanzetti*.

vehicular homicide the unintentional killing of another during a motor vehicle accident. See *homicide*. Compare with *involuntary manslaughter*.

vendetta a series of attacks motivated by revenge. Vendettas generally are reprisals by organized criminals for a real or perceived wrong.

vengeance the human desire to get even for a real or perceived injustice. Vengeance finds expression in the criminal law as *retribution*, one of several justifications for punishment.

venue the location where an offender is to be tried for a crime. See *change of venue*.

Vera Institute of Justice a nonprofit organization in New York City dedicated to improving the administration of justice through innovative research and practice.

verdict a decision by the jury or panel of judges at the conclusion of a trial. See *jury*.

V

vertical prosecution   a prosecutorial strategy where the prosecutor remains with the same case from the beginning to its conclusion. Vertical prosecution is thought to have advantages over *horizontal prosecution*.

veterans court   a *specialty court* that processes veterans with substance abuse and/or mental health problems.

VICAP   Violent Criminal Apprehension Program (VICAP), an initiative of the *Federal Bureau of Investigation (FBI)*. VICAP consists of a computerized database contributed by local law enforcement agencies of detailed information on homicides, sexual assaults, and other violent crimes. The VICAP computer is programmed to make daily searches for patterns that can identify possible crime series, which enables federal, state, and local investigators to avoid *linkage blindness*. The practical effectiveness of VICAP as a national investigatory tool for violent crime was compromised by the unwillingness of local law enforcement agencies to regularly contribute necessary data. See *National Center for the Analysis of Violent Crime (NCAVC)*.

vicarious liability   a legal responsibility one has for the behavior of specific others, such as their children or employees. Compare with *strict liability*.

vicarious traumatization   the trauma suffered by individuals who work closely with victims of crime and other traumatized populations. Symptoms of vicarious traumatization can include social withdrawal, sleep difficulties, and aggression.

Vice Lords   a large Chicago-based *street gang*.

vices   moral violations, such as *prostitution*, *gambling*, or drugs. Vices are regarded by many as *victimless crimes*.

vice squad   a police detective unit that specializes in investigating prostitution, gambling, drugs, and other so-called *vices*.

victim   an individual, business, or organization that suffers harm or loss as a result of a crime. See *secondary victim*.

victim advocate   a social worker who assists victims of crime to obtain services and to navigate their way through the criminal justice process. Victim advocates, who may be former crime victims themselves, refer victims to counseling, accompany them to court hearings, and sometimes participate in the preparation of victim impact statements. See *victim assistance program*, *victim impact statement*.

victim assistance program   a program, often affiliated with a prosecutor's office, that refers crime victims to counseling and other needed services, and helps them navigate their way through the criminal justice process.

Victim assistance programs, which are staffed by victim advocates, can also be independent nonprofit agencies. See *victim advocate.*

victim compensation program   the formalized practice of paying a monetary sum by government to crime victims, most often those who have suffered as a result of a violent offense, such as murder or physical or sexual assault. Victim compensation programs consider claims from victims who have filed an application, generally compensating victims of violent crimes. Federal funding for such programs comes in part from the *Office for Victims of Crime (OVC).*

victim impact statement   a written or oral statement prepared and delivered prior to sentencing to inform a criminal court of the physical, emotional, financial and other effects that a crime had on the victim, the family, friends, and the community. Victim impact statements are also prepared for parole boards for the purpose of offering information about the offender's suitability for parole. See *allocution.*

victimization   the result of a crime in terms of human suffering or property loss.

victimization rate   the amount of criminal victimization within a given population measured as a proportion of people victimized per unit of population.

victimization survey   a survey of citizens designed to determine the incidence and correlates of criminal victimization. Victimization surveys became popular with the recognition of the shortcomings of traditional crime statistics, such as the *Uniform Crime Reports (UCR).* See *National Crime Victimization Survey (NCVS), International Crime Victimization Survey (ICVS).*

victimless crime   a type of crime where there is no identifiable victim who has suffered harm or loss. *Prostitution* is often considered to be a victimless crime because both the prostitute and the *john,* or client, receive what they want: money, drugs, or some other gain to the prostitute, and sex for the client. However, the victim of such a crime can be a third party, such as residents who live in a *red light district* and see a corresponding drop in their property values, or a less measurable sense of moral decay within the surrounding community. More serious crimes can sometimes accompany victimless crimes, such as armed robbery of *johns* by prostitutes. See *overcriminalization.*

victim notification   communication by state or local authorities that an offender is having a hearing or is about to be released. Victim notification is just one aspect of *victims' rights.*

victim-offender mediation   efforts to bring together offender and victims to help each understand the perspectives and difficulties of the other. See *victim-offender reconciliation program (VORP).*

V

victim-offender reconciliation program (VORP)   a program designed to bring victims and offenders together in an effort to gain mutual understanding and healing.

victimology   the field of study that focuses on victims of crime and related issues.

victim-precipitated crime   a crime in which the victim plays a role in instigating the offender into a violent act. *Marvin E. Wolfgang* elaborated on this concept in his study of criminal *homicide*. Critics of this concept maintain that it is the offender who engages in the wrongful behavior and therefore is responsible for their actions.

victims' bill of rights   a piece of legislation that ensures crime victims' rights, such as the right to offer a *victim impact statement* in court and the right to notification of a convicted offender's release from prison.

victims' rights   a broad term that conveys concern for the ability of victims to more fully participate in the criminal justice system and to avoid further frustration and pain.

victim witness program   see *victim assistance program*.

video piracy   the illegal reproduction, distribution, or sale of unauthorized copies of commercial videotapes.

Vidocq Society   a Philadelphia-based organization of former police officers, scientists, and other specialists who selectively reanalyze unsolved criminal cases in hopes of solving them. Membership in the Vidocq Society is not open to the public.

vigilante   one who engages in *vigilantism*.

vigilantism   the illegal practice by citizens of assuming the roles of police, prosecutor, judge, and sometimes executioner in a misguided attempt to seek justice.

vintage fraud   the sale of *counterfeit* wine.

violence   physical aggression that results, or has the potential to result, in injury.

Violence Against Women Act   a federal law originally passed in 1994 to protect female victims and provide funding for programming.

violent crime   crimes against persons that involve actual or potential bodily harm. Examples of violent crime include *murder, rape, robbery,* and *aggravated assault*. Compare with *property crime, public order crime*.

Violent Crime Burden Index   a measure of the incidence of crime for various jurisdictions in India. Compare with *Uniform Crime Reports*.

V

Violent Crime Impact Team an initiative of the federal *Bureau of Alcohol, Tobacco, Firearms and Explosives (ATF)* to combat firearms related violent crime by targeting the worst offenders and their organizations.

violent video games video games that depict violent situations and frequently encourage players to participate by responding with violence. Research has shown that violent video games can increase aggression in players.

visiting room a room in a *prison*, *jail*, or other correctional facility where inmates can meet with visitors.

visual criminology the analysis of the visual images related to crime and punishment in the media and other outlets. Visual criminology is interested in how visual images in film and various other media shape and distort thinking and behavior.

vital statistics public health data on births and deaths.

voice stress analysis an electronic means of analyzing voice patterns to detect deception.

voir dire the process of selecting jurors prior to the commencement of a criminal trial. During voir dire, prospective jurors are questioned by both the *prosecutor* and the *defense attorney* to learn something about their backgrounds and possible biases. See *jury selection*, *peremptory challenge*.

Vold, George (1896–1967) an American sociologist and criminologist of the 20th century whose works included *Theoretical Criminology*. Vold spent most of his career at the University of Minnesota.

Volstead Act the federal legislation which prohibited the manufacture and sale of alcoholic beverages. The Volstead Act prompted the beginning of *Prohibition*.

volume crime crimes that have the most impact on a community. Compare with *victimless crime*.

voluntary manslaughter an intentional killing under circumstances that mitigate the seriousness down from murder because the killing lacks the element of malice necessary to be found guilty of *murder*. Compare with *involuntary manslaughter*.

volunteers in probation (VIP) a program in which lay volunteers assist probation officers in performing their duties and in meeting the needs of those on probation.

voyeurism the practice of viewing people or materials where they are depicted, often with little or no clothing, for the purpose of sexual gratification.

waiver the transfer of juvenile cases to adult court. Waivers are generally used by officials in the case of serious crimes, such as *murder*.

Walker spy ring an American spy ring that consisted of John Walker, his son, his brother, and his son's friend. The Walker spy ring passed information to the Soviets from the 1960s to the 1980s. It was determined that the information relayed by the Walker spy ring resulted in the deaths of American operatives. See *espionage*.

Walnut Street Jail a jail of the late 18th century in Philadelphia where prisoners spent their time alone in their cells. The Walnut Street Jail was illustrative of the *Pennsylvania system*.

Walsh, Adam (1974–1981) a boy who was abducted from a Florida shopping mall and later murdered. See *abduction, Walsh, John*.

Walsh, John (1945—present) the father of *Adam Walsh* who, after the murder of his son, became involved in victims' rights. John Walsh eventually served as host of the television show *America's Most Wanted*.

wanted felon a felon sought by authorities.

wanted poster a paper notice announcing that a fugitive is wanted and providing identifying details and recent whereabouts.

war crimes crimes committed during war that violate international law. Examples of war crimes include the rape of civilians, summary executions, and torture. War crimes often are tried in the International Criminal Court in The Hague, The Netherlands.

war criminal an offender who commits *war crimes*. Examples include the Nazis who were convicted at the Nuremberg trials. See *human rights violations*.

warden the administrator of a *prison* or other secure correctional facility.

warning shot the discharge of a firearm as a signal to stop. Law enforcement officers are advised against using warning shots because the bullet could accidentally injure or kill someone and the officer has one less round for self-defense.

W

war on crime   euphemism to describe government's large-scale effort to combat criminal behavior.

war on drugs   term used to describe the attempt by federal and state authorities to control the supply and distribution of illegal drugs. The war on drugs has drawn criticism for its emphasis on *supply reduction* and is likened by many to *Prohibition,* which failed in its efforts to stem the flow of illegal alcohol. Compare with *demand reduction.* See *drug czar, Office of National Drug Control Policy (ONDCP).*

warrant   a legal order issued by a *judge* that permits law enforcement authorities to arrest a *suspect* or search a specific location.

warrantless search   a search of a person, vehicle, or dwelling without a *search warrant.* Warrantless searches may be justified under certain circumstances.

warrantless surveillance   the surveillance of citizens by the government purportedly to gather intelligence about *terrorist* activities.

Warren Commission   the federal commission chaired by Chief Justice Earl Warren to investigate the *assassination* of President John F. Kennedy. The Warren Commission, formally The President's Commission on the Assassination of President Kennedy, concluded that the assassination was the work of *Lee Harvey Oswald* operating alone. Many have since disputed these findings, some proposing that Kennedy's assassination was the culmination of a vast conspiracy, variously attributed to *organized crime,* communists, or the *Central Intelligence Agency (CIA).*

Warren Court   the United States Supreme Court under the leadership of Chief Justice Earl Warren. The Warren Court became known for several landmark decisions affecting criminal justice.

wartime trade violations   the prohibited sale or exchange of specific goods reserved for the war effort.

Washington Navy Yard Shooting   the September 16, 2013, killing of 18 victims and the wounding of three others at the Washington Navy Yard in Washington, D.C. The perpetrator, Aaron Alexis, was shot and killed by police.

watch system   an early system of policing where watchmen protected the walls and gates of the city, and kept order among citizens. In contrast to modern police, those keeping peace under the watch system did not wear uniforms or carry firearms, but instead rang bells at regular intervals.

waterboarding   the controversial practice of placing an individual on a horizontal or inclined board and pouring water over the mouth and nose, creating a sensation of drowning for the purpose of eliciting information during *interrogation*. See *enhanced interrogation techniques*, *Guantanamo Bay*, *torture*.

Watergate the 1972 break-in at the Democratic National Committee headquarters at the Watergate complex in Washington, D. C. and the ensuing attempts by the Richard M. Nixon administration to cover it up. The Watergate scandal led to the *indictment* and *conviction* of several conspirators and eventually to the resignation of President Nixon himself. See *burglary*, *political crime*.

WAVR-21 a 21-item instrument designed to assess the risk of workplace and campus violence. The WAVR-21 was developed by psychologists Stephen White and J. Reid Meloy. See *workplace violence*.

wear pattern physical and distinguishing features of frequently used objects, such as shoes or tires that can help investigators solve crimes.

Weather Underground a radical left-wing organization founded in 1969 whose goal was to overthrow the U.S. government. They were responsible for the bombing of government buildings and banks.

Wecht, Cyril (1931–present) an American forensic pathologist whose career and involvement in highly publicized criminal cases has included controversy. Wecht spent much of his career as the medical examiner and coroner of Allegheny County, Pennsylvania.

Wedtech Corporation a New York business and defense contractor that bilked the U.S. government out of millions of dollars in the 1980s.

Weed and Seed a program initiated by the U.S. Department of Justice where communities were given funds and technical assistance to "weed" out criminal elements and "seed" neighborhoods with programs and social support.

welfare fraud intentional misuse of state **welfare** systems by withholding information or giving false or inaccurate information in order to obtain welfare benefits.

Westgate Massacre the *terrorist* attack on the Westgate shopping mall by *al-Shabaab* in Nairobi, Kenya on September 21, 2013 that left 67 dead and 175 wounded. See *massacre*.

Wetterling Act federal legislation that mandates that convicted sex offenders must register with local law enforcement agencies in the communities where they live. Failure of sex offenders to comply with the provisions of the Wetterling Act can result in subsequent felony charges.

whipping post a vertical post where persons were fastened to undergo whipping as a punishment.

W

whistle-blower   an individual who reports or exposes misconduct or alleged dishonest or illegal activity occurring in an organization. The discovery of *corporate crime* and other types of *organizational deviance* often depend on someone willing to serve as a whistle-blower. Despite federal and state laws intended to protect them, whistle-blowers often face threats and other reprisals. See *qui tam suit*.

White Aryan Resistance   a neo-Nazi separatist organization based in Warsaw, Indiana. Compare with *Aryan Nations*.

white-collar crime   crimes committed by persons of high social status in the course of their occupations. White-collar crimes often involve a betrayal of trust. The term was coined by American criminologist *Edwin H. Sutherland*. Compare with *corporate crime, occupational crime, occupational deviance*.

white slavery   the illegal practice of selling or trading humans, often young females, generally for the purpose of forcing them to enter into *prostitution*. Compare with *human trafficking, sex trade*. See *Mann Act*.

white widow   a potent strain of *cannabis sativa* developed in the Netherlands.

Whitman, Charles (1941–1966)   a former Marine marksman who in 1966 shot and killed 18 people and wounded 30 others on the campus of the University of Texas at Austin, after killing his wife and mother. Whitman, who was an engineering student at the university, was shot and killed by police when they stormed the library tower where he positioned himself to wreak this terror. See *mass murder*.

whodunit   a *crime novel* where the murder *suspect* is not readily known. Whodunits often have several possible suspects. Generally the identity of the *perpetrator* is not revealed until the end of the novel. Compare with *mystery, suspense novel*.

whorl   a circle-like segment of a human *fingerprint*. Compare with *arch, loop*.

Wickersham Commission   a federal commission convened in the 1930s to study problems in law enforcement and the administration of justice. Compare with *National Advisory Commission on Criminal Justice Standards and Goals*.

wife beating   the *physical assault* of a wife by her spouse. See *spousal abuse, domestic violence*. Compare with *husband beating*.

wilderness camp   a camp in a remote wooded setting designed as a correctional alternative for *delinquent* youths. Compare with *boot camp*.

wildlife crime   offenses against wildlife, such as *poaching*. An example of wildlife crime is the killing of African elephants for their ivory tusks.

W

Williams, Wayne (1958–present) a *serial killer* believed to be responsible for the deaths of 28 young Black men and boys in and around the city of Atlanta, Georgia in the early 1980s. Williams lured his victims with promises of fame. When the victims agreed to participate in homosexual acts, Williams killed his young victims and dumped their bodies in various locations around Atlanta. He was sentenced to life imprisonment. See *serial murder*.

Wilson, James Q. (1931–2012) a political scientist and criminologist best known for his conservative recommendations on crime control and his criticisms of criminology. For much of his career, Wilson was affiliated with Harvard University. His published works include *Thinking about Crime*.

wire fraud *fraud* accomplished via interstate electronic wire communications.

wiretapping the practice of intercepting telephone transmissions for the purpose of gathering evidence for possible criminal prosecution. Wiretapping became an issue in the wake of the terrorist bombings of the World Trade Center and the Pentagon when it was recognized that traditional wiretapping laws may no longer be sufficient to maintain national security against such threats.

withdrawal the physical and psychological condition brought on by the discontinuation of drugs of *addiction*. In some cases, withdrawal can result in the death of the addicted patient. See *methadone maintenance, substance abuse.*

without prejudice a designation of criminal case dismissals indicating that the case can again be pursued. Compare with *with prejudice.*

with prejudice a designation of criminal case dismissals indicating that the case cannot be refiled. Compare with *without prejudice.*

witness a person who has personal, firsthand knowledge of a crime through the use of one or more of their senses, most often sight. Compare with *eyewitness.*

witness identification the process of identifying a criminal suspect by someone who has first-hand knowledge through the use of their senses.

witness intimidation the act of threatening witnesses with harm. If successful, witness intimidation can result in dismissal of charges against the defendant. See *Witness Security Program.*

witness protection program see *Witness Security Program.*

Witness Security Program a federal program that permits witnesses testifying in dangerous cases to establish new identities in new locales to protect them and their families from possible reprisals. Those who take advantage

W

of this program typically are those who provide critical testimony in criminal cases against *organized crime*. The Witness Security Program is administered by the U.S. Marshals Service.

witness stand   a platform with a low wall behind which witnesses sit and testify in a courtroom. In western courtrooms, the witness stand is located adjacent to the bench to the judge's left. See *witness*.

Wolfgang, Marvin E. (1924–1998)   an American criminologist who spent most of his career at the University of Pennsylvania. Wolfgang is best known for pioneering studies of criminal homicide, the assessment of crime seriousness, and measurement of delinquency over the life course. He was a student of criminologist *Thorsten Sellin* with whom he worked for much of his career. Wolfgang served as president of the *American Society of Criminology* and worked with prominent criminologists around the world, including *Franco Ferracuti* and *Sir Leon Radzinowicz*. See *Pennsylvania School of Criminology*.

Wootton, Barbara (1897–1988)   a British sociologist and criminologist. Her works include *Social Science and Social Pathology* and *Crime and Criminal Law*.

workhouse   a local or regional correctional facility designed to house inmates for less than one year, and where inmates work as part of the correctional process. Compare with *jail*.

workplace violence   violent acts by employees or by their relatives and acquaintances occurring at or around places of employment. The problem of workplace violence, which gained national attention with shooting rampages by postal employees, has been the subject of specialized prevention strategies. See *WAVR-21*.

work release   a conditional release of an inmate from a correctional facility for the purpose of maintaining employment, obtaining training or education, or other rehabilitative purposes. Those on work release must return to the facility at the end of the day.

World Trade Center bombing   the 1993 *terrorist* bombing of the World Trade Center that killed six people and injured numerous others. Compare with *Attack on America*.

World Society of Victimology (WSV)   an international organization whose purpose is to promote research in *victimology* and improve practices related to crime victims.

wound ballistics   the study of the effect of projectiles on the human body. See *wound morphology*.

wound morphology   the study of the size, shape, and structure of wounds.

wound track   the separation of tissue and bone caused by a weapon as it enters the human body.

Wournos, Aileen (1956–2002)   a female *serial killer* who murdered several men in Florida in the 1980s. Wuornos, a prostitute, shot and killed men who thought they were going to have sex with her. She claimed to suffer from a history of abuse. She was executed in 2002. See *serial murder*.

wrongful conviction   the conviction in court of an accused person who, in fact, did not commit the alleged offense. A wrongful conviction can occur when a defendant, out of fear of a more severe *sentence*, may plead guilty to a crime they did not commit. It can also occur as the result of erroneous *eyewitness identification*. The advent of *DNA testing* in criminal cases has led to the exoneration of a number of persons who were wrongfully convicted in the past. See *convicted innocent, Innocence Project*.

wrongful execution   the execution of a person who did not commit the crime for which the penalty of death was eventually imposed. See *convicted innocent, wrongful conviction*.

wrongful prosecution   the prosecution of an individual who did not commit the crime in question. Compare with *wrongful conviction*.

XYY chromosome   see *extra Y chromosome.*

Yakuza   a Mafia-like criminal organization in Japan. See *organized crime*.

yard   a secure outside area where prison inmates gather and recreate.

Youngstown Gang   a notorious group of professional criminals operating in Youngstown, Ohio in the 1960s and 1970s. They committed a series of robberies, burglaries, and various other property crimes, all with the knowledge and support of not only *organized crime* in the area, but also local law enforcement.

youth court   a program that teaches young people to serve as judges, prosecutors, defense attorneys, and juries in actual cases involving their peers.

Youth for Justice   a program of the *American Bar Association* designed to reduce juvenile *delinquency* through *law-related education*.

Youth Handgun Safety Act   a federal law prohibiting those under 18 years of age from possessing handguns and the sale or transfer of handguns.

Youth Risk Behavior Survey (YRBS)   a periodic survey of high school youths about their dietary habits, substance abuse, fighting, and other behaviors which pose a risk to health. The YRBS is sponsored by the *Centers for Disease Control and Prevention (CDC)*.

youth service bureau   an organization that offers prevention and intervention services for youths, especially those at risk of entering the juvenile or criminal justice system.

Zapruder film an amateur movie film taken by Abraham Zapruder of the 1963 *assassination* of President John F. Kennedy. The Zapruder film was used in subsequent investigations. See *Oswald, Lee Harvey*.

Zebra killings a series of racially motived homicides in California in the 1970s. Four African American men were convicted of the slayings and sentenced to life imprisonment. See *multiple homicide*.

zero tolerance a policy where officials do not tolerate any offending behavior, such as weapons in school and public order crimes. Zero tolerance policies by law enforcement were inspired in part by the *broken windows theory.*

Zetas a powerful Mexican *drug cartel* responsible for hundreds of murders.

Zimmerman, George (1983–present) an American who shot and killed Trayvon Martin in Sanford, Florida. Zimmerman claimed he thought his life was in danger when he shot Martin. See *stand your ground laws*.

Zionist Occupation Government (ZOG) name given to the U.S. government by certain anti-Semitic, right-wing extremist groups. The name is intended to convey that the government is controlled by Jews.

Zodiac killer a *serial killer* who murdered seven people and taunted authorities in northern California in the late 1960s and early 1970s. The killings were never solved.

zoophilia sexual relations between a human and an animal. Also called *bestiality*.

$SAGE researchmethods

The essential online tool for researchers from the world's leading methods publisher

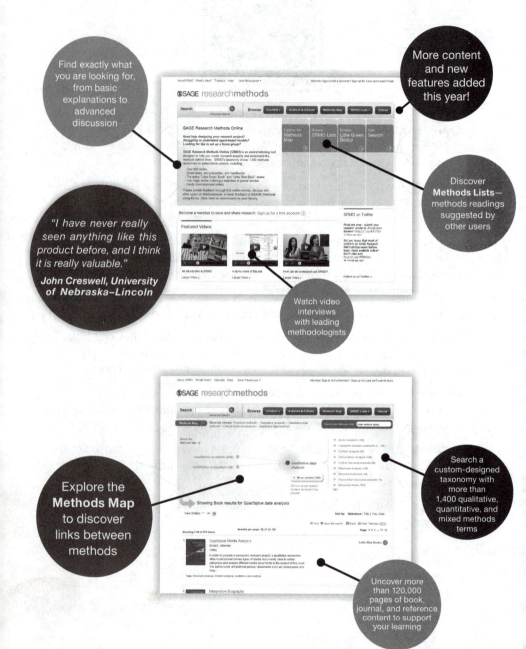

Find exactly what you are looking for, from basic explanations to advanced discussion

More content and new features added this year!

"I have never really seen anything like this product before, and I think it is really valuable."
John Creswell, University of Nebraska–Lincoln

Discover Methods Lists—methods readings suggested by other users

Watch video interviews with leading methodologists

Explore the Methods Map to discover links between methods

Search a custom-designed taxonomy with more than 1,400 qualitative, quantitative, and mixed methods terms

Uncover more than 120,000 pages of book, journal, and reference content to support your learning

Find out more at
www.sageresearchmethods.com